Action Mathematics 5

Doris Bass and Ann Farnham

Cassell

CASSELL LTD.
1 St. Anne's Road, Eastbourne
East Sussex BN21 3UN

© Cassell & Co. Ltd. 1971
All rights reserved. No part of this publication may
be reproduced, stored in a retrieval system, or transmitted
in any form or by any means, electronic, mechanical,
photocopying, recording or otherwise, without the prior
permission in writing of the Publishers.

First published 1972
Second impression November 1972
Third impression May 1974
Fourth impression June 1975
Fifth impression June 1978
Sixth impression January 1980
Seventh impression 1984

Design and Graphics by Peter McClure

ISBN 0 304 93803 3

Printed in Hong Kong by Colocraft Ltd.

Contents

Pie Charts	1
Fractions, Decimal Fractions, Percentages	4
Percentages	5
Following Instructions	6
Cash Accounts	7
Imprest Petty Cash Accounts	8
Line Graphs	10
Drawing Parallel Lines	12
A Property of Right-Angled Triangles	13
Using Pythagoras' Theorem	14
Degree of Accuracy	16
Approximation	19
Facts from Diagrams	20
Rates	21
Combined Transformations	22
Volume	24
Constructions	27
Problems Using Sets	28
Formulae	32
Indices	34
Logarithmic Curve	35
Using Logarithmic Curves	37
Logarithms	40
Using a Logarithm Table	41
Polyhedra	45
Tessellation	46
Income Tax	48
Drawing to Scale	50
Probability	52
Pascal's Triangle	58
Probability with Dice	59
A Tour of the Wye Valley	60
Savings	62
Family Tree	64
Ancestry of a Drone Bee	65
Using Isometric Paper	66
Reasoning	67
Compound Interest	68
Depreciation	69
The Nephroid	70
The Cardioid	72
The Spiral	73
Logarithm Tables	74
Useful Information	76

The **Action Mathematics** series consists of five books. Answers are available separately.

The logarithm tables are reprinted by permission of Cambridge University Press from *Elementary Tables* by the S.M.P.

Pie Charts

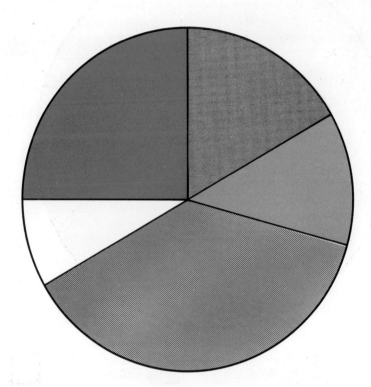

(1) Measure each of the angles at the centre of the circle.
(2) What fraction of the whole circle is coloured (a) red (b) grey (c) blue (d) violet (e) white?
(3) The diagram represents the total number of pullovers sold by a retailer during a given period. If he sold 240 pullovers during the period, find the number of each colour sold.

B

A grower advertised a bargain parcel which contained 30 daffodil, 27 tulip, 45 snowdrop, 18 hyacinth and 20 narcissus bulbs and 40 crocus corms.

(1) Find the total number of bulbs and corms contained in the parcel.
(2) Write down the fraction of the total number which were:
    (a) daffodils   (b) tulips   (c) snowdrops   (d) hyacinths   (e) narcissi
    (f) crocuses.
(3) Make a carefully labelled pie chart showing the information given in the advertisement.

C

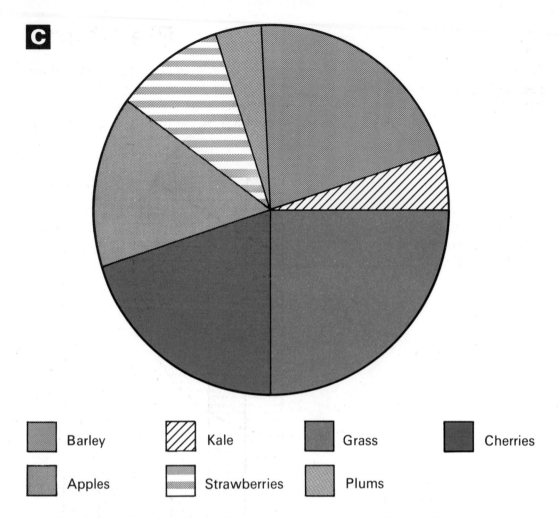

(1) The pie chart shows how the land is cultivated on Mayday Farm. Each sector represents the area of land used for growing a particular crop. Measure the angles at the centre of the circle carefully. Copy and complete this table. One item has been completed.

Crop	Angle at centre of sector	Fraction of total area used for the crop
Barley	75°	$\frac{5}{24}$

(2) The total area under cultivation at Mayday Farm is 120 hectares (ha). Find the area used in growing each crop.

D

The total area under cultivation on a small farm is 90 hectares. The area used for growing each crop is as follows:
Barley 45 ha, Grass 4½ ha, Rye 13½ ha, Wheat 9 ha, Sugar beet 18 ha.
Draw a large circle and make a pie chart showing how the total area of cultivated land is used.

E

The column graph shows the area under cultivation at Orchard Farm.

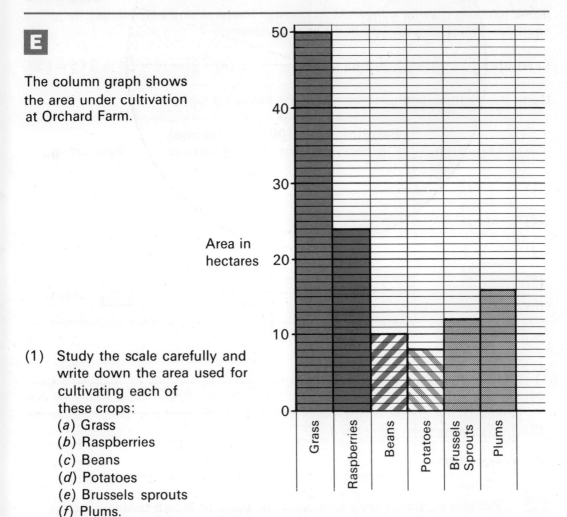

(1) Study the scale carefully and write down the area used for cultivating each of these crops:
   (a) Grass
   (b) Raspberries
   (c) Beans
   (d) Potatoes
   (e) Brussels sprouts
   (f) Plums.
(2) What is the total area under cultivation?
(3) What fraction of the total area is used for each crop?
(4) Draw a circle with radius 5 cm and make a pie chart showing how the total area of the farm is cultivated.
(5) When is it possible to draw a column graph to show data which could not be shown on a pie chart?

3

Fractions, Decimal Fractions, Percentages

A

Some fractions may be converted to decimal fractions and percentages by giving equivalent fractions with 100 in the denominator.

Example (i) $\frac{17}{25} = \frac{68}{100} = 0\cdot 68 = 68\%$ (ii) $\frac{39}{60} = \frac{13}{20} = \frac{65}{100} = 0\cdot 65 = 65\%$

Using this method, copy and complete the following table.

	Fraction	Fraction with 100 in denominator	Decimal fraction	Percentage
(1)	$\frac{19}{20}$			
(2)	$\frac{13}{20}$			
(3)	$\frac{3}{5}$			
(4)	$\frac{18}{25}$			
(5)	$\frac{23}{25}$			
(6)	$\frac{36}{60} = \frac{*}{20}$			
(7)	$\frac{69}{75} = \frac{*}{25}$			
(8)	$\frac{26}{40} = \frac{*}{20}$			
(9)	$\frac{102}{120} = \frac{*}{20}$			
(10)	$\frac{117}{180} = \frac{*}{20}$			

B

Another method, suitable for some fractions, is shown below.

(i) $\frac{1}{2} = \frac{50}{100} = 0\cdot 50 = 50\%$ (ii) $\frac{1}{4} = \frac{1}{2}$ of $\frac{1}{2} = 0\cdot 25 = 25\%$

(iii) $\frac{1}{8} = \frac{1}{2}$ of $\frac{1}{4} = 0\cdot 125 = 12\frac{1}{2}\%$

Write each of the following as a (*a*) decimal (*b*) percentage.

(1) $\frac{3}{4}$ (2) $\frac{3}{8}$ (3) $\frac{5}{8}$ (4) $\frac{7}{8}$ (5) $\frac{1}{16}$

Percentages

A number of four-year-old cars were tested and the results tabulated as follows.

Type	Tested	Failed
A	150	42
B	320	128
C	180	54
D	480	84
E	72	27
F	256	64

(1) Find the percentage failure for each type of car.
(2) Make a column graph showing the percentage failures in ascending order.
(3) Does this result *necessarily* reflect on the cars? Give reasons for your answer.

A large number of householders were questioned as to whether they possessed a refrigerator. The results were tabulated.

Copy and complete this table.

Town	Number of householders questioned	Number owning a refrigerator	Percentage of those questioned who owned refrigerators
G	8,000	6,400	
H	6,000	2,220	
J	7,500	2,100	
K	15,000	2,550	
L	21,000	11,970	
M	3,800	1,140	

Make a column graph showing in descending order the percentage of householders in each town owning a refrigerator.

Following Instructions

(1) Draw a square ABCD with sides 5 cm long. Draw the diagonals AC and BD. Mark the centre of the square with the letter O. From C draw, inside the square, the longest possible arc of a circle with radius OC.
Shade the part of the square which touches the arc but is not inside any part of the circle.

(2) Draw a rectangle PQRS so that PQ and RS are both 8 cm long and PS and QR are 5 cm long.
With P as centre and radius equal to PS draw the longest possible arc of a circle inside the rectangle.
Draw a line parallel to PQ and at a distance of 3 cm from PQ.
Mark a point which is 5 cm from P and also 3 cm from PQ.

(3) Draw an equilateral triangle with sides 6 cm long.
Using each side in turn as diameter draw semi-circles outside the triangle.
Join the mid-points of the three sides to make a small triangle.
Using each side of the small triangle in turn as a diameter draw semi-circles outside the small triangle.
Shade the parts of the diagram which are not inside any semi-circle.

(4) Draw a circle centre O with radius 4 cm.
Draw two diameters AB and CD at right-angles to one another.
Using AO as diameter draw a circle. Repeat this using OB, OC and OD in turn as diameters.
Shade with one colour the parts of the diagram which are inside 3 circles.
Shade with another colour the parts which are only inside one circle.

(5) Give clear and short instructions for drawing this figure.

Cash Accounts

Use paper ruled for cash accounts. If this is not available prepare the paper. Complete each of the following cash accounts and balance on the suggested date.

A

May 8	Cash in hand £65·40; Paid C. May £3·58
May 9	Sold goods for cash £38·63; Paid fares £0·84
May 10	Bought goods for cash £49·42; F. Mann paid his account £6·50 Sold goods for cash £17·34
May 11	Sold goods for cash £24·68
May 12	Paid wages £22·50; Paid rent £21·60; Sold goods for cash £35·17

Balance on 12 May.

B

Sept. 18	Cash in hand £43·36; Cash sales £24·17; Paid D. Smith's a/c £9·76
Sept. 19	Cash sales £28·35; Paid carriage on goods £2·45
Sept. 20	Cash purchases £51·76; Cash sales £25·47; Paid fares 87p
Sept. 21	Bought stamps 96p; Paid insurance £2·68; Paid wages £21·50
Sept. 22	Cash sales £27·84

Balance on 22 September.

C

Feb. 28	Cash in hand £103·20; Paid rent £26·50; Paid L. Grey £28·19; Cash purchases £94·83
Feb. 29	Bought stamps 125p; Cash sales £32·85
March 1	Cash purchases £57·18; Paid fares £1·30; Cash sales £32·07
March 2	Cash sales £56·58; Paid P. Bowen's account £19·55
March 3	Paid window cleaner 85p; Paid shop cleaner £4·20; Cash sales £41·75

Balance on 3 March.

Imprest Petty Cash Accounts

Jane kept the Petty Cash Account for her firm. She was told the **imprest amount** was £10. This meant that she was given £10 at the beginning of the week. During the week she spent £7·35 and had a balance of £2·65 at the end of the week. She kept this balance and was given £7·35 so that she started the second week with £10. During the second week she spent £9·15 and had a balance of £0·85 which she kept. She was given £9·15 so started the third week with £10. In this way she started each week with £10. This is the first week's account. Copy it carefully and carry it on for the next two weeks. The items are entered as for a Cash Account.

		£			£
1972			1972		
Oct. 2		10·00	Oct. 2	Postage	1·47
			2	Fares	2·68
			3	Telegram	0·54
			4	Stationery	1·28
			5	Fares	0·35
			6	Postage	1·03
				Total	7·35
			6	Balance c/d	2·65
		£10·00			£10·00
Oct. 9	Balance b/d	2·65			
	Cash	7·35			

At any time during the week the money spent plus the balance in hand should be equal to the imprest amount.

Note: A cashier who keeps a cash account should always obtain a receipt for any money spent.
When money for fares, etc., is given to a messenger he should be asked to sign a petty cash voucher.

Second week's expenditure

Carry this on after the account you have copied.

Oct. 9 | Stationery £0·86; Postage £1·18
10 | Fares £1·47; Typewriter ribbons £1·25
11 | Book £0·96; Postage £0·58
12 | Window cleaning £0·85
13 | Postage £1·37; Fares £0·63

Third week's expenditure

Carry on directly after the second week.

Oct. 16 | Cash book £1·25; Postage £0·72
17 | Ink £0·35; Fares £0·67
18 | Cleaning materials £1·18; Postage £1·28
19 | Telegrams £0·96; Postage £0·89; Tips £0·30
20 | Carriage on goods £0·47; Fares £0·18

---

(1) How much did the cashier give to the petty cash account at the beginning of the fourth week?

(2) How much was spent on fares during the three weeks?

(3) How much during the three weeks was spent on items dealt with by the Post Office? (Include telegrams.)

Line Graphs

Line graphs are not always **straight** lines. Some points when joined form a curve.

An example of this is when a number is graphed against its square, when $y = x^2$. We choose a value for x and square it to obtain y.

x = 0	1	2	3	4	5	6	7
↓	↓	↓	↓	↓	↓	↓	↓
y = 0	1	4	9	16	25	36	49

The line graph $y = x^2$ is shown below.

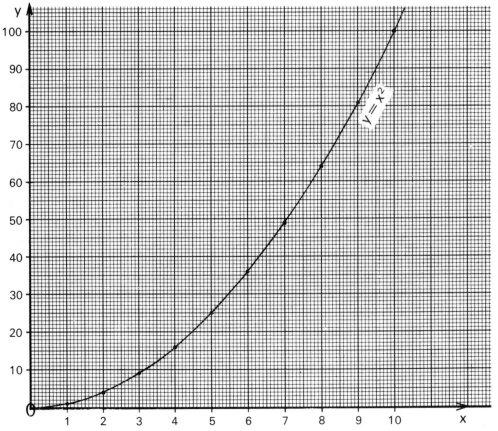

You will notice that different scales are used on the x and y axes. Can you suggest a reason for this?

Note that 1 mm on the x-axis represents 0·1 unit while on the y-axis 1 mm represents 1 unit.

Use graph paper. Draw rectangular axes and draw the graph of $y = x^2$ for as many values of x as your paper allows.

A

To find 6.8^2, correct to the nearest unit.

At point A on the curve $x = 6.8$. The y co-ordinate at this point is 46.

$\rightarrow 6.8^2 = 46$, correct to the nearest unit

Find the square, correct to the nearest unit, of each of the following.

(1) 4·5
(2) 8·5
(3) 2·3
(4) 7·6
(5) 3·4

(6) 6·1
(7) 5·7
(8) 9·8
(9) 1·4
(10) 3·9

B

To find $\sqrt{23}$, correct to 1 decimal place.

At point B on the curve, $y = 23$. The x co-ordinate at this point is 4·8.

$\rightarrow \sqrt{23} = 4.8$ correct to 1 decimal place

Find the square root, correct to 1 decimal place, of each of the following.

(1) 20
(2) 54
(3) 75
(4) 96
(5) 67

(6) 19
(7) 85
(8) 46
(9) 72
(10) 8

C

Find, to the degree of accuracy indicated:

(1) $\sqrt{83}$ correct to 1 decimal place
(2) 4.9^2 correct to nearest unit
(3) 8.1^2 correct to nearest unit
(4) $\sqrt{92}$ correct to 1 decimal place
(5) $\sqrt{24}$ correct to 1 decimal place
(6) 2.6^2 correct to nearest unit

Drawing Parallel Lines

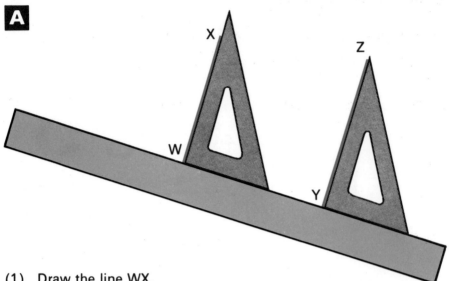

A

(1) Draw the line WX.
(2) Place the edge of a set square along the line WX as shown.
(3) Place the ruler along the edge of the set square which is perpendicular to the line WX.
(4) Keep the ruler steady, and slide the set square to a new position, keeping it against the edge of the ruler. Draw the line YZ.
  WX and YZ are parallel.

B

Draw these figures to the given measurements.
The two short sides are parallel. The two long sides are parallel. Each figure is a parallelogram.

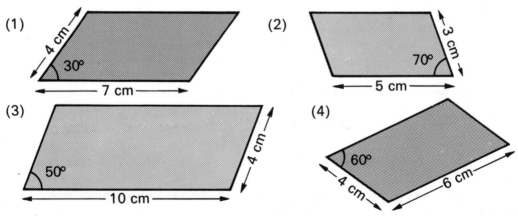

A Property of Right-Angled Triangles

Pythagoras demonstrated the truth of the statement that 'In any right-angled triangle, the square on the hypotenuse (the side opposite the right angle) is equal to the sum of the squares on the other two sides', by using the geometrical methods of his time.

You can show the statement to be true by a 'jigsaw' or dissection method.

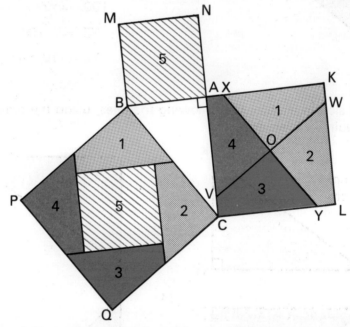

(1) Draw any right-angled triangle without measuring the sides. Label the right angle A, so that BC is the hypotenuse. Make AC longer than AB.
(2) Draw the squares on the three sides.
(3) Find the centre of the square ACLK. Label it O.
(4) Divide the square ACLK as shown, with one line XY parallel to BC, and the other VW perpendicular to BC. These lines must pass through O, the centre of the square.
(5) Cut out the squares ABMN and ACLK after you have coloured and numbered the 4 sections into which ACLK has been divided. Cut the square ACLK into the 4 sections.
(6) Fit the 5 pieces into the square BCQP.
You will find that
the square on AB + the square on AC = the square on BC.
$$AB^2 \quad + \quad AC^2 \quad = \quad BC^2$$

Using Pythagoras' Theorem

Example

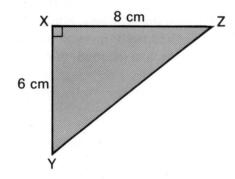

To find YZ.

$XY^2 + XZ^2 = YZ^2$
$6^2 + 8^2 = YZ^2$
$36 + 64 = YZ^2$
$100 = YZ^2$
$\rightarrow \quad YZ = \sqrt{100}$
$YZ = 10$ cm

Find the third side in each of the following triangles, using the graph on page 10 to find the required square root.

(1) $KM = 3$ cm
$ML = 4$ cm
Find LK.

(2) $PQ = 7$ cm
$PR = 5$ cm
Find QR.

(3) $SU = 8$ cm
$TU = 4$ cm
Find ST.

(4) $WX = 6$ cm
$XY = 6$ cm
Find WY.

(5) This is the plan of a room. Find the shortest distance from corner G to corner K.

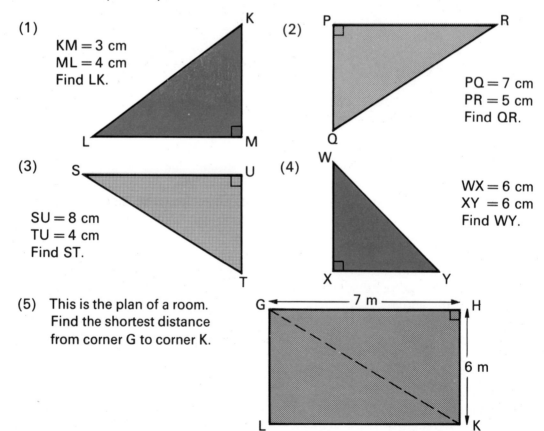

Practical Uses

(1)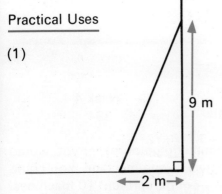

The first floor window of a house is 9 m above the ground, which is horizontal. The flower bed is 2 m wide.
What length of ladder is needed to reach the window from just beyond the flower bed?

(2)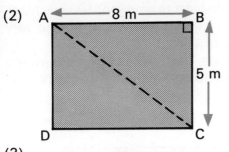

The diagram represents the floor of a room. Find the length of cable needed
(a) to go from A to C via B;
(b) to go from A to C by the most direct route.

(3)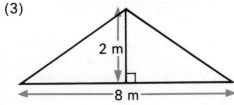

The diagram shows the rafters of a house. How long are the rafters?

(4)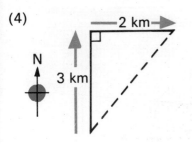

John walks 3 km due north, and then 2 km due east. Find the shortest distance he could walk to return to his starting point.

(5)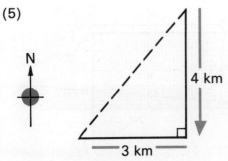

Jane walks 4 km due south and then 3 km due west. If she returns to her starting point by the most direct route, find the total distance she has walked.

15

Degree of Accuracy

Sales of Machinery

Week 1	Week 2	Week 3	Week 4
572	457	508	394

If you were asked to make a column graph on centimetre graph paper you would realise at once that you could not use 1 cm to represent one machine. You could not, without having a very long strip of paper, use 1 cm to represent 10 machines. You would come to the conclusion that 1 cm would have to represent 100 machines and would understand that the only graph you could draw would not be very accurate though it would give a reasonable idea of sales.

Using a scale of 1 cm to represent 100 machines, each 0·1 cm (1 mm) would represent 10 machines and you would need to consider the figures 'to the nearest ten'.

Week 1   572 is greater than 570 and less than 580 but is nearer 570 so you would represent this by 5·7 cm.

Week 2   457 is greater than 450 and less than 460 but is nearer 460 so you would represent this by 4·6 cm.

Week 3   508 is greater than 500 and less than 510 but is nearer 510 so you would represent this by 5·1 cm.

Week 4   394 is greater than 390 and less than 400 but is nearer 390 so you would represent this by 3·9 cm.

The graph would look like this.

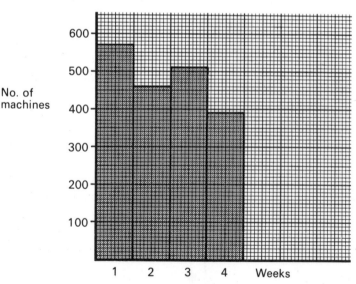

The graph shows that having started well sales have dropped in the fourth week.

A

Factory:	A	B	C	D	E	F	G
No. of employees:	461	487	506	284	316	397	344

The table shown above gives the number of employees at each of 7 factories. You are required to draw a column graph on centimetre graph paper giving the approximate number employed in each factory. You will need to work to a scale of 1 cm representing 100 people employed.

(1) Copy and complete this table.

	A	B	C	D	E	F	G
No. of employees to the nearest ten	460						
Distance representing this number on the required scale	4·6 cm						

(2) Draw the required graph.
(3) Write down (i) a question to which answers obtained from your graph would be sufficiently accurate (ii) a question to which answers obtained from your graph would not be sufficiently accurate.

B

A survey showed that the population in each of 8 villages on a given date was as follows:

P 841   T 803
Q 768   U 928
R 537   V 865
S 1154   W 758

(1) Arrange these figures in descending order.
(2) Draw a column graph to show to the nearest ten the population of the villages in descending order.

Note Village V: The number is midway between 860 and 870 so you may use either of these numbers.

The approximate height of each of ten Swiss ski-resorts is given as follows:

Adelboden	1353 metres		Kanderstag	1167 metres
Braunwald	1249 metres		Leysun	1298 metres
Davos	1556 metres		Murren	1645 metres
Engelberg	1049 metres		St Moritz	1827 metres
Grindelwald	1055 metres		Zermatt	1617 metres

Make a column graph showing these heights correct to the nearest 100 metres.

The attendance at a football match was 9,674. For numbers as large as this the scale of a graph drawn on centimetre graph paper could be 1 cm representing 1000 people and 0·1 cm (1 mm) represents 100 people. The numbers would have to be corrected to the nearest hundred people before the graph could be drawn.

9,674 is between 9600 and 9700 but nearer 9700.

On the suggested scale the column would be 9·7 cm in height.

(1) Copy and complete this table.

Ground	Actual attendance	Attendance to nearest 100	Height of column
A	9,674	9,700	9·7 cm
B	4,730		
C	10,018		
D	3,671		
E	7,270		
F	9,483		
G	12,617		
H	6,346		
J	3,654		
K	5,825		

(2) What difference has the adjustment made to the figures shown as the total attendance at the 10 matches?

(3) Draw a column graph showing as accurately as the scale allows the attendance at each of the matches.

(4) What statements about attendance at these matches would be justified from consideration of your graph?

Approximation

When measuring the length of a road there would be little point in giving the answer to the nearest millimetre. But in precision engineering a measurement to 0·1 mm could be required. When measuring or giving answers a sensible degree of accuracy is needed.

In each of the following exercises give the measurements or amounts to the degree of accuracy stated.

Write to the nearest whole number in the units given.
- (1) 5·67p
- (2) 28·37p
- (3) 54·83 cm
- (4) 7·613 km
- (5) £4·07
- (6) £241·55
- (7) 11·76 litres
- (8) 4·37 cm
- (9) 0·89 km
- (10) 87·48 m
- (11) £99·75
- (12) 10·34 litres

B

Write each of the following correct to one decimal place.
- (1) 2·54 cm
- (2) 6·38 m
- (3) 14·47 g
- (4) 9·81 m
- (5) 48·61 m
- (6) 8·438 km
- (7) 16·48 m
- (8) 0·71 g
- (9) 147·06 km
- (10) 28·37 g
- (11) 5·08 m
- (12) 10·76 km

Write each of the following correct to two decimal places.
- (1) £4·3874
- (2) £16·1807
- (3) £5·8436
- (4) £28·2626
- (5) 32·1234 km
- (6) £19·0079
- (7) 6·2176 km
- (8) £110·9249
- (9) 316·584 km
- (10) £4·0583
- (11) 17·6197 km
- (12) £41·0086

Write each of the following correct to 3 significant figures.
- (1) £63·47
- (2) £2·2367
- (3) 496·3 km
- (4) 128·8 km
- (5) 9·476 cm
- (6) £2867
- (7) £1473
- (8) £16·786
- (9) 4·059 m
- (10) 5·043 km
- (11) £8·327
- (12) £0·6787

Facts from Diagrams

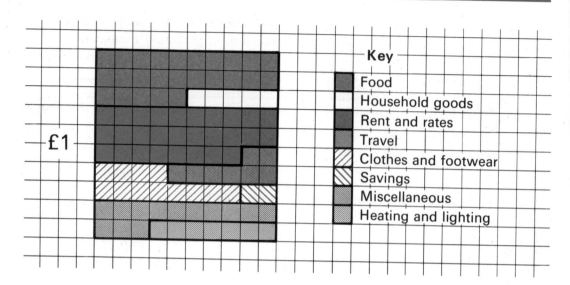

A survey in a certain area gives a pattern of personal expenditure as shown in the diagram.
This is a picture of personal spending as an average for this area.
Each small square represents 1p. The questions refer to people living in this area.

(1) How much was spent on each item?
(2) Check your result by seeing if the total is £1.
(3) On this estimate, how much does a man earning £18* per week spend on
   (a) food
   (b) rent and rates
   (c) travel
   in one week?
(4) How much would a man earning £32* per week spend on
   (a) food
   (b) clothing and footwear
   (c) household goods
   in one week?
(5) On an income of £1200* per year, how much would a man spend per year on
   (a) travel
   (b) miscellaneous items
   (c) savings?

*These figures give the actual amount paid after Income Tax, Insurance, etc., have been deducted.

Rates

The local rate in Hillston is 70p in the £.

The rateable value of four houses and a bungalow are given below.

W	X	Y	Z	Z
£60	£140	£110	£90	£90

The ratepayer living in each house pays 70p in every £ of the rateable value.
The occupier of house Z pays (90 × 70)p = 6300p
 = £63

(1) Find the rates paid each year for each of the houses and the bungalow.
(2) How much is paid to the Borough Council as rates for these 5 houses in one year?

The local rate in Woodhampton is 65p in the £.
(1) Find the rate paid by each tenant of houses of the same rateable value as the above, but situated in Woodhampton.
(2) How much does the Woodhampton Borough Council receive in rates from the 5 houses in one year?

Copy and complete:

		Rateable value	Townside Local rate 80p in the £	Newstead Local rate 55p in the £	Larchville Local rate 62p in the £
(1)	House P	£90			
(2)	House Q	£105			
(3)	House R	£108			
(4)	House S	£128			
(5)	House T	£156			

Combined Transformations

The following strip patterns have been made by combining some of the transformations you have already studied. The shape is transformed; the colouring is used only to make the pattern attractive.

Draw and cut out an interesting shape from thin card.

(1) **Translation**
This pattern shows a repeated translation of the shape.

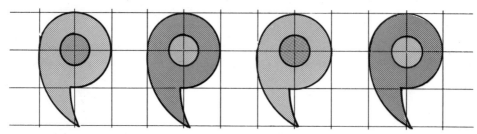

(2) **Reflection**

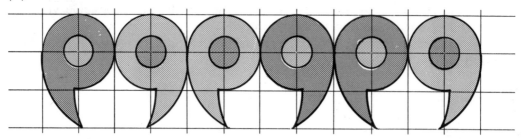

(3) **Translation followed by a half turn**

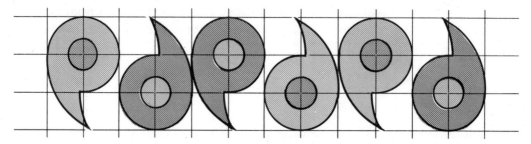

(4) **Translation followed by a reflection**
Each of these patterns shows a translation followed by a reflection in the broken line. Why are the patterns different?

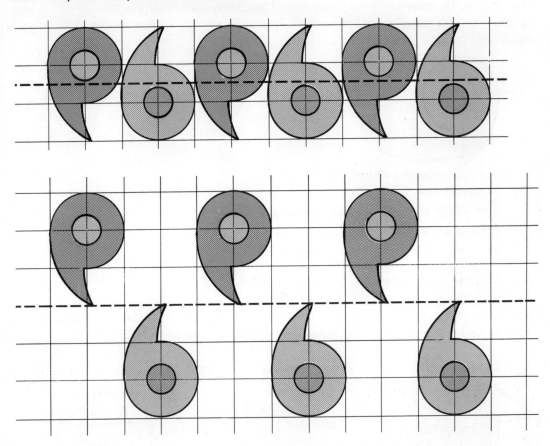

(5) **Reflection followed by a half turn**

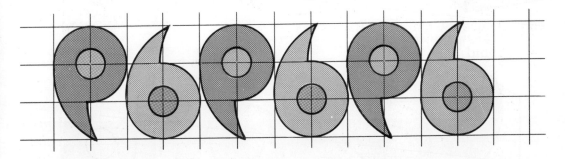

Do any of these transformations give the same pattern?

23

Volume

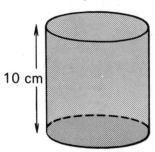
Cylinder
10 cm
Radius of base = 3 cm

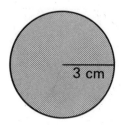
Circular base of cylinder
3 cm

Volume of cylinder   = Area of base × height
Area of circular base = π × 3 × 3 cm²
                     =   3·14 × 9 cm²
                     = 28·26 cm²
Volume of cylinder   = 28·26 × 10 cm³
                     = 282·6 cm³

Find the volume of each of the following solids.

(1) **Cube**

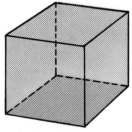

Edge of cube = 5 cm

(2) **Cylinder**

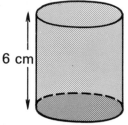

6 cm

Radius of base = 2 cm

(3) **Triangular based prism**

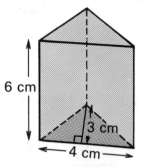

6 cm
3 cm
4 cm

(4) **Cuboid**

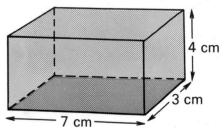

4 cm
3 cm
7 cm

(5) **Cylinder**

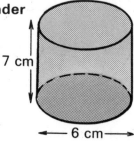

7 cm
6 cm

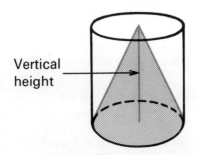

A cone may be cut from a cylinder as shown in the diagram. The vertical height of the cone is the same as that of the cylinder. How much material do you think is wasted?

A rectangular based pyramid may be cut from a cuboid as shown in the diagram. The vertical height of the pyramid is the same as that of the cuboid. How much material do you think is wasted?

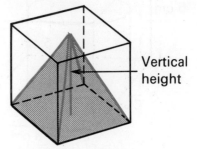

Did you guess that half or a quarter of the material is wasted? In fact, $\frac{2}{3}$ is wasted in each case.
The volume of the cone is $\frac{1}{3}$ of the volume of the cylinder.
The volume of the pyramid is $\frac{1}{3}$ of the volume of the cuboid.

Find the volume of each of the following cones.
In each case the vertical height of the cone is the same as that of the cylinder from which it could be cut.

(1)   (2)   (3)

Vol. of cylinder = 12 cm³   Vol. of cylinder = 18 cm³   Vol. of cylinder = 22 cm³

Find the volume of each of the following pyramids.
In each case the vertical height of the pyramid is the same as that of the cube or cuboid from which it could be cut.

(1) 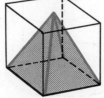 (2)  (3)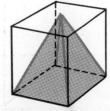

Vol. of cuboid = 24 cm³   Vol. of cuboid = 21 cm³   Vol. of cuboid = 20 cm³

Example

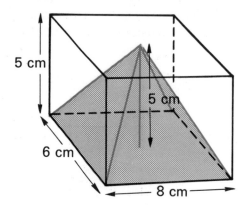

Volume of cuboid = Area of base × height

Area of base of cuboid = 8 × 6 cm²
= 48 cm²

Volume of cuboid = 48 × 5 cm³
= 240 cm³

Volume of pyramid = $\frac{1}{3}$ × 240 cm³
= 80 cm³

C

Find the volume of each of the following cones or pyramids.

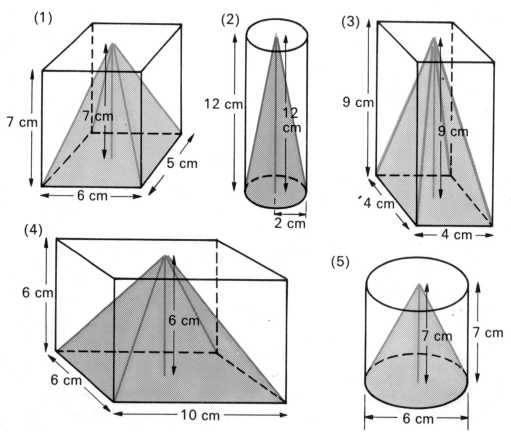

Constructions

The diagrams on this page have been drawn to the scale of 1 cm representing 4 cm.

(*a*) Make any measurements you think necessary and draw each of the diagrams to the scale 1 cm representing 2 cm.
(*b*) Name each of your diagrams.

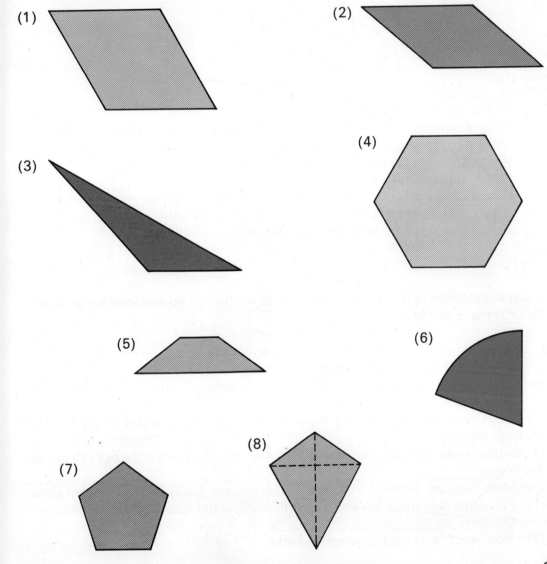

27

Problems Using Sets

A

A group of 20 students take the Secretarial course.
  Set X are those who take typewriting.   Set Z are those who take accounts.
  Set Y are those who take shorthand.
The initials of the students are shown in this Venn diagram:

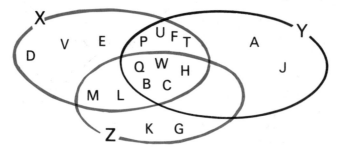

(1) Write down the initials of the students in (i) Set X (ii) Set Y (iii) Set Z.
(2) Make a list of students who take all three subjects.
(3) Which students take shorthand and typewriting but not accounts?
(4) Which students take shorthand and typewriting?
(5) Which students take typewriting and accounts but not shorthand?
(6) How many students take shorthand and accounts only?
(7) How many students take only one subject?

B

A party of children was taken to a fair and for each child an entrance fee of 4p was paid. Of these children,
  28 went on the Big Dipper,
  31 went on the Dodgem Cars,
  28 rode on the Scenic Railway.
10 children went on both the Big Dipper and Dodgem Cars but not on the Scenic Railway.
6 children went on both the Big Dipper and the Scenic Railway but not on the Dodgem Cars.
11 children rode on both the Scenic Railway and Dodgem Cars but not on the Big Dipper.
9 children went on all three—the Big Dipper, Scenic Railway and Dodgem Cars.
(1) From this data make a Venn diagram and find the total number of children in the party.
(2) How much was paid in entrance fees?

X = {owners of TV sets} = {John, Ann, Vera, Fay, Charles, Bill, Michael}

Y = {owners of record players} = {Philip, Mary, David, Ann, Clare, John, Keith}

Z = {owners of tape recorders} = {Peter, Vera, David, Ben, Ann, Don, Frank, John}

(1) Draw a large Venn diagram similar to that on the opposite page.
(2) Write the name of each student in the correct region.
(3) How many students were questioned?
(4) How many students owned all three items?
(5) How many students owned two items?
(6) How many students owned just one item?
(7) How many students had at least two of the items?

A group of boys and girls live in a village and belong to a Youth Club in a nearby town. The club has many members and various sections but we are only considering the members who live in the village and belong to one or more of three sections.

Don, Emma, Ron, Philip, Anne, Marie and Martin belong to the Sports Club (Set S).

Tim, Jean, Elaine, Yvonne, Ron, Philip, George and Marie belong to the Cycling Club (Set C).

Yvonne, Elaine, Jean, Ron, Jim and Philip are in the Hockey Team (Set H).

Make a large Venn diagram showing the intersection of the three sets. Put each member's name in the correct region. Use your diagram to help in answering these questions:

(1) A hockey match was arranged to take place on the same day as a cycling expedition. Which members had to choose between the two events?
(2) The secretary of the Cycling Club was a girl who took care not to arrange a cycling expedition on the same day as a sports meeting. Who was the secretary?
(3) Don, the treasurer, collected 28p from each member for fares to and from the town on the club night. All the village members attended. How much did Don collect?

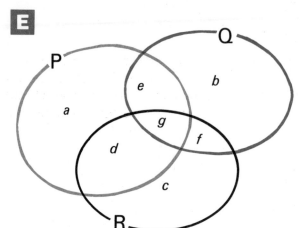

P = {shopkeepers who stock Lite}

Q = {shopkeepers who stock Brite}

R = {shopkeepers who stock Flite}

The diagram represents three sets of shopkeepers who stock washing powders. Some sell just one type of powder, some stock two brands and some stock all three.
Each of the regions in the diagram is labelled with a small letter.
The region *g* represents shopkeepers who stock all three brands.
We write: *g* represents $P \cap Q \cap R$.
The region *e* represents shopkeepers who stock Lite and Brite but do not stock Flite. We write: $P \cap Q \cap R'$.
The region *a* represents shopkeepers who stock only Lite. We write: $P \cap Q' \cap R'$.

Describe the shopkeepers represented by each of the regions *b, c, d* and *f*. Write each in terms of intersection of sets.

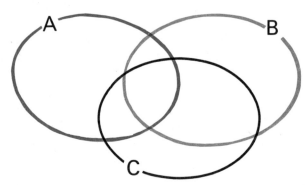

Copy this Venn diagram 5 times. Shade the region which represents each of these sets.
Diagram 1 $A \cap B \cap C$
Diagram 2 $A \cap B \cap C'$
Diagram 3 $B \cap C \cap A'$
Diagram 4 $A \cap B' \cap C'$
Diagram 5 $A \cap C$

G

An English examination was arranged in three parts A, B and C. A master with 30 candidates gave each a number from 1 to 30 so that names did not appear on the answer papers. When he had finished marking the master realised that he had not marked a total of 90 papers. He drew the following Venn diagram to illustrate the situation.

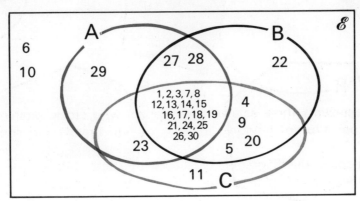

Use the Venn diagram and answer the following questions.
(1) How many candidates completed:
    (a) all three papers,   (b) only two papers,   (c) only one paper?
(2) How many candidates were absent from all the examinations?
(3) Which candidates did not complete paper A?
Think hard before you answer the next question.
(4) How many examination papers did the master mark?

H

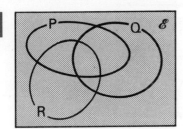

The Venn diagram represents a universal set $\mathscr{E}$ and three subsets P, Q and R.
$\mathscr{E}$ = {whole numbers 2 to 20}
P = {odd numbers between 2 and 20}
Q = {numbers divisible by 3 between 2 and 20}
R = {numbers divisible by 5, from 1 to 20}

Make a list of each of the three sets P, Q and R.
Copy the Venn diagram and put each number in the appropriate region.
Write down the members of each of the following sets and describe each set in a short sentence.
(1) P∩Q∩R   (2) P∩Q∩R'   (3) P∩Q
(4) Q∩R   (5) P∩R'∩Q'
(6) Which numbers are in P or in Q or in R? We write this as set P∪Q∪R.
(7) Which numbers are not in any of the sets P, Q or R? We write this as set (P∪Q∪R)'.

Formulae

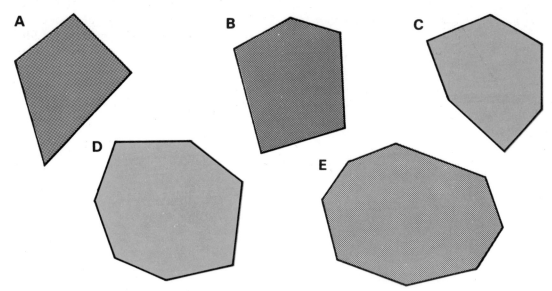

Trace each of the polygons drawn above, and find how many diagonals can be drawn in each polygon.

Copy and complete:

Polygon	Number of diagonals
A	
B	
C	
D	
E	

There is a formula which can be used so that the figures need not be drawn each time.

Number of diagonals $= \frac{1}{2}n(n-3)$, where n is the number of sides of the polygon.

Check your answers by using this formula:
e.g. C $n = 6$
Number of diagonals $= \frac{1}{2} \times 6 \times 3$
$= 9$

Formulae are often useful. Each of the following questions should be answered by applying the formula given.

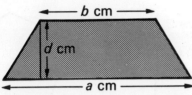

Area of trapezium $= \dfrac{(a+b)}{2} \times d$ cm²

Find the area of the trapezium when (1) $a=4$, $b=2$, $d=3$
(2) $a=9$, $b=3$, $d=4$
(3) $a=10$, $b=5$, $d=6$

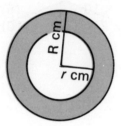

Area of ring $= \pi(R+r)(R-r)$ cm²
$\pi = 3.14$

Find the area of the ring when (1) $R=6$, $r=4$
(2) $R=2$, $r=1$
(3) $R=5$, $r=3$

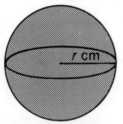

Volume of sphere $= \tfrac{4}{3}\pi r^3$ cm³
$\pi = 3.14$

Find the volume of the sphere when (1) $r=1$
(2) $r=3$

D

A temperature of C° Centigrade can be converted to F° Fahrenheit by the formula
$$F = 32 + \tfrac{9}{5}C$$
Find F, when   (1) $C=5$ (2) $C=10$ (3) $C=15$

E

A temperature F° Fahrenheit can be converted to C° Centigrade by the formula
$$C = \tfrac{5}{9}(F-32)$$
Find C, when   (1) $F=104$ (2) $F=95$ (3) $F=32$

33

Indices

A

Example (a) $a^2 \times a^3 = (a \times a) \times (a \times a \times a) = a^5$
(b) $a^2 \times a^3 = a^{(2+3)} \qquad = a^5$

Using both methods (a) and (b) give each of the following products in index form.

(1) $a^3 \times a^1$
(2) $a^4 \times a^2$
(3) $x^5 \times x^2$
(4) $y^2 \times y^2$
(5) $a^4 \times a^3$
(6) $10^2 \times 10^3$
(7) $10^3 \times 10^1$
(8) $2^3 \times 2^4$
(9) $5^4 \times 5^1$
(10) $10^3 \times 10^3$

B

Example (c) $a^5 \div a^3 = \dfrac{\overset{1}{\cancel{a}} \times \overset{1}{\cancel{a}} \times \overset{1}{\cancel{a}} \times a \times a}{\underset{1}{\cancel{a}} \times \underset{1}{\cancel{a}} \times \underset{1}{\cancel{a}}} = a^2$

(d) $a^5 \div a^3 = a^{(5-3)} \qquad = a^2$

Using both methods (c) and (d), give each of the following quotients in index form.

(1) $a^3 \div a^1$
(2) $x^7 \div x^4$
(3) $y^6 \div y^2$
(4) $m^5 \div m^4$
(5) $x^6 \div x^3$
(6) $10^5 \div 10^3$
(7) $5^7 \div 5^3$
(8) $4^6 \div 4^5$
(9) $10^8 \div 10^5$
(10) $10^4 \div 10^2$

C

You know that
$10^1 = 10, \qquad 10^2 = 100, \qquad 10^3 = 1000, \qquad 10^4 = 10{,}000$

Do you think it true or false to say that $10^0 = 0$?

You probably thought it true, but consider this:

$$10^2 \div 10^2 = \dfrac{\overset{1}{\cancel{10}} \times \overset{1}{\cancel{10}}}{\underset{1}{\cancel{10}} \times \underset{1}{\cancel{10}}} = 1$$

$$10^2 \div 10^2 = 10^{(2-2)} = 10^0$$

But the two answers must have the same result so 10^0 must be equal to 1.

What is the value of (a) 5^0 (b) 7^0 (c) x^0 (d) y^0 (e) 12^0?

Logarithmic Curve

Draw the graph of $y = 10^x$ when $x = 0, 1, 2$ and 3.
To do this the value of y for each value of x must be found.

When $x = 0$, $y = 10^0 \rightarrow y = 1$ at A.
When $x = 1$, $y = 10^1 \rightarrow y = 10$ at B.
When $x = 2$, $y = 10^2 \rightarrow y = 100$ at C.
When $x = 3$, $y = 10^3 \rightarrow y = 1000$ at D.

These points, when plotted, cannot be joined by a straight line. They must be joined with a smooth curve.
You will realise that the scale on the y-axis is very small compared with the scale on the x-axis.

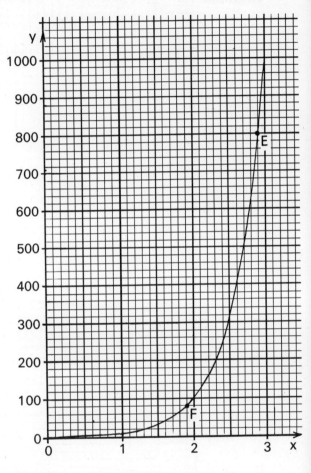

What is the value of x when $y = 800$? This is shown at E where

$y = 800$, $x = 2 \cdot 9$ (approx.) $\rightarrow 800 = 10^{2 \cdot 9}$ (approx.)

When $y = 80$ at F, $x = 1 \cdot 9$ (approx.) $\rightarrow$ $80 = 10^{1 \cdot 9}$ (approx.)

Do not attempt to draw the graph, use the graph given above to write each of the following numbers to base 10. The answers will only be accurate to the nearest one place of decimals.

(1) 60   (2) 400   (3) 200   (4) 260
(5) 160   (6) 540   (7) 660   (8) 20

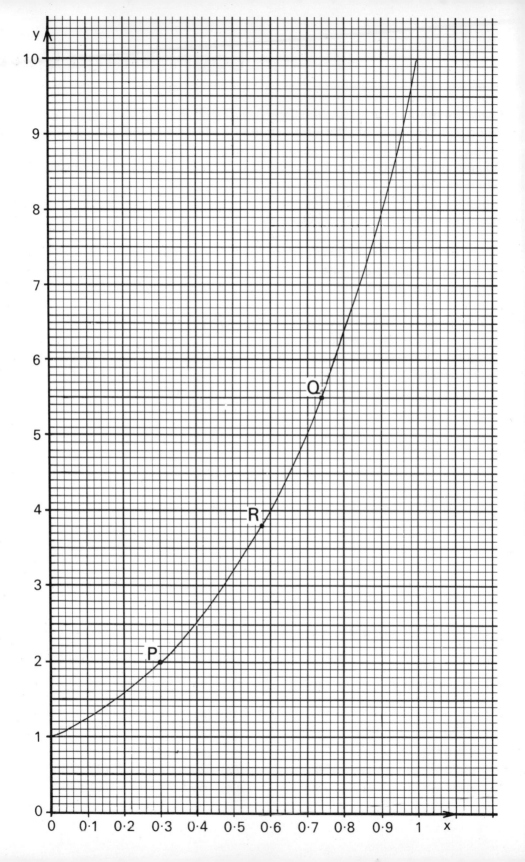

Using Logarithmic Curves

The graph on page 35 is drawn on too small a scale for reading values of x between 0 and 1, and values of y between 0 and 10. The graph on the opposite page will enable you to read these values approximately.
Do not attempt to draw the graph. Use the one printed on the opposite page to estimate the required value of x.

Example At P, $y = 2$, $x = 0.3 \to 2 = 10^{0.3}$
At Q, $y = 5.5$, $x = 0.74 \to 5 = 10^{0.74}$.

Copy and complete each of the following tables, giving answers in **A** correct to one place of decimals and those in **B** correct to two decimal places.

A

	y	x	y = 10ˣ
(1)	2	0.3	$2 = 10^{0.3}$
(2)	2.5		
(3)	4		
(4)	5.1		
(5)	8		
(6)	1.6		
(7)	3.2		
(8)	6.3		

B

	y	x	y = 10ˣ
(1)	4.2	0.62	$4.2 = 10^{0.62}$
(2)	1.8		
(3)	2.9		
(4)	3.5		
(5)	5.7		
(6)	9.1		
(7)	8.4		
(8)	7.6		

It is possible from the graph to find y when x is known.

Example At R, $x = 0.58$, $y = 10^{0.58} \to y = 3.8$

C

Find as accurately as is possible the value of y in each of the following examples.

(1) $x = 0.42$
(2) $x = 0.59$
(3) $x = 0.73$
(4) $x = 0.83$
(5) $x = 0.88$
(6) $x = 0.23$
(7) $x = 0.08$
(8) $x = 0.95$
(9) $x = 0.82$
(10) $x = 0.74$

Write down the value of y when

(1) x = 1·73   (2) x = 2·08   (3) x = 1·59

(4) x = 2·23   (5) x = 2·95   (6) x = 1·83

From the readings you have made from the curve on page 36 you can find approximate values of x greater than 1 for values of y greater than 10.

Example I $2 \cdot 5 = 10^{0 \cdot 4}$
$25 = (2 \cdot 5 \times 10) = 10^{0 \cdot 4} \times 10^1 = 10^{(0 \cdot 4 + 1 \cdot 0)} = 10^{1 \cdot 4}$
$250 = (2 \cdot 5 \times 100) = 10^{0 \cdot 4} \times 10^2 = 10^{(0 \cdot 4 + 2 \cdot 0)} = 10^{2 \cdot 4}$

Example II $4 \cdot 2 = 10^{0 \cdot 62}$
$42 = 10^{1 \cdot 62}$
$420 = 10^{2 \cdot 62}$

Use the readings you made in previous exercises and write each of the following numbers to base 10 as accurately as possible.

	(a)	(b)	(c)
(1)	1·8	18	180
(2)	5·7	57	570
(3)	4	40	400
(4)	6·3	63	630
(5)	2·9	29	290
(6)	6·6	66	660
(7)	9·1	91	910
(8)	8·4	84	840

The curves drawn from the formula $y = 10^x$ are called **logarithmic** curves. The word 'logarithm' comes from two Greek words, 'logos' meaning ratio and 'arithmos' number.

The two words 'Arithmos logos'—ratio of numbers—were the title of a work by Archimedes, a mathematician who lived from 287 to 212 B.C.

You have seen that $10^2 \times 10^3 = 10^{(2+3)} = 10^5 = 100{,}000$
and that $10^6 \div 10^4 = 10^{(6-4)} = 10^2 = 100$

It is possible to multiply and divide any numbers using this method and remembering that the degree of accuracy is limited.

Example I $2 \times 4 = 10^{0 \cdot 3} \times 10^{0 \cdot 6} = 10^{(0 \cdot 3 + 0 \cdot 6)} = 10^{0 \cdot 9}$
When $x = 0 \cdot 9$ then $y = 8$
This is the answer you would expect. Quite rightly you are thinking that this is a long and clumsy method of multiplying 2 by 4. But you might be happy to use this method to find the product of 5·7 and 7·6.

Example II $5 \cdot 7 \times 7 \cdot 6 = 10^{0 \cdot 76} \times 10^{0 \cdot 88} = 10^{(0 \cdot 76 + 0 \cdot 88)}$
$= 10^{1 \cdot 64}$
When $x = 1 \cdot 64$, $y = 43 \rightarrow 5 \cdot 7 \times 7 \cdot 6 = 43$ (correct to 2 figs.)

Use the logarithmic curve on page 36 to complete each of the following calculations. The answers should be correct to 2 figs.

(1) 5·1 × 1·6
(2) 3·2 × 1·8
(3) 1·6 × 4·2
(4) 5·8 × 2·9
(5) 7·6 × 3·2

(6) 6·3 × 9·1
(7) 4·2 × 3·5
(8) 57 × 1·8
(9) 29 × 8·4
(10) 3·5 × 3·7

Example III $7 \cdot 6 \div 3 \cdot 2$
$= 10^{0 \cdot 88} \div 10^{0 \cdot 50} = 10^{(0 \cdot 88 - 0 \cdot 50)} = 10^{0 \cdot 38}$
When $x = 0 \cdot 38$, $y = 2 \cdot 4 \rightarrow 7 \cdot 6 \div 3 \cdot 2 = 2 \cdot 4$ (correct to 2 figs.)

Use the logarithmic curve to complete each of the following calculations correct to 2 figures.

(1) 9·1 ÷ 2·9
(2) 8·4 ÷ 1·6
(3) 2·9 ÷ 1·8
(4) 3·5 ÷ 2·9
(5) 8·4 ÷ 3·5

(6) 63 ÷ 35
(7) 84 ÷ 76
(8) 5·7 ÷ 2·9
(9) 57 ÷ 18
(10) 63 ÷ 9·1

Logarithms

Work on indices and series led to an invention in Mathematics. John Napier, born in the sixteenth century, worked hard for twenty years before producing a small book on logarithms and their use. This invention has been of great value and saved hours which would have been spent in making calculations. After Napier's death his friend Henry Briggs compiled logarithm tables such as we use today.

The principles are the same as those you have already used but a greater degree of accuracy can be obtained from using tables than can be obtained from using a graph.

Logarithmic tables have been made to various degrees of accuracy. We shall use three-figure tables.

From the graph you can find that $45 = 10^{1 \cdot 65}$

$1 \cdot 65$ is called the **logarithm** of 45.

In this case '1'—the number before the decimal point—is called the **characteristic**.

The number after the decimal point, in this case $\cdot 65$, is called the **mantissa**.

Using a Logarithm Table

Below is an extract from a logarithm table giving answers to 3 figures.

P	Section Q									
	0	1	2	3	4	5	6	7	8	9
4·0	0·602	603	604	605	606	607	609	610	611	612
4·1	0·613	614	615	616	617	618	619	620	621	622
4·2	0·623	624	625	626	627	628	629	630	631	632
4·3	0·633	634	635	636	637	638	639	640	641	642
4·4	0·643	644	645	646	647	648	649	650	651	652
4·5	0·653	654	655	656	657	658	659	660	661	662
4·6	0·663	664	665	666	667	667	668	669	670	671
4·7	0·672	673	674	675	676	677	678	679	679	680
4·8	0·681	682	683	684	685	686	687	688	688	689
4·9	0·690	691	692	693	694	695	695	696	697	698

To find the logarithm of 4·76:
- (*a*) The number is greater than 0 and less than 10 so the characteristic is 0.
- (*b*) Look down column **P** and find the first two numbers 4·7. Lay your ruler horizontally across the table just below the figures 4·7.
- (*c*) Look along this line until you come to the figure under the column headed 6 in Section **Q**. This number, ·678, is the mantissa you need.

The logarithm of 4·76 is 0·678 so you can write
$$4·76 = 10^{0·678}$$

You will notice that the table gives logarithms of numbers which are greater than 0 and less than 10.

For numbers greater than 10 and less than 100, the number has been multiplied by 10^1 and the characteristic is 1.
Example $47·6 = 4·76 \times 10^1 = 10^{0·678} \times 10^1 = 10^{1·678}$

For numbers greater than 100 and less than 1000 the number has been multiplied by 10^2 and the characteristic is 2.
Example $476 = 4·76 \times 10^2 = 10^{0·678} \times 10^2 = 10^{2·678}$

Use the extract given above to find the logarithm of each of the following numbers and write each number to base 10.

(1) 4·37   (2) 4·58   (3) 4·12   (4) 4·96   (5) 4·05
(6) 44·3   (7) 48·1   (8) 47   (9) 424   (10) 465

Calculations can be made using logarithms but the answers must be given as ordinary numbers.

You will remember that when you were told that $y = 10^{0.58}$ you found, from the graph, that $y = 3.8$.

This value of y can be found quite easily using a logarithm table.

Example $y = 10^{0.685}$, the logarithm of y is 0·685.

Using the extract given on the opposite page, find the mantissa ·685. Lay your ruler horizontally across the table under ·685. Look in column **P** on this line and you find 4·8. Then look at the number at the top of Section **Q** in the same column as ·685. You find 4 which is the third figure in your answer.

When $y = 10^{0.685}$ then $y = 4.84$
When $y = 10^{1.685}$ then $y = 4.84 \times 10 = 48.4$
When $y = 10^{2.685}$ then $y = 4.84 \times 100 = 484$

Use the extract from the table to find y when

(1) $y = 10^{0.614}$
(2) $y = 10^{0.669}$
(3) $y = 10^{0.690}$
(4) $y = 10^{0.639}$
(5) $y = 10^{1.651}$

(6) $y = 10^{1.603}$
(7) $y = 10^{1.642}$
(8) $y = 10^{2.673}$
(9) $y = 10^{2.684}$
(10) $y = 10^{0.652}$

You will find a logarithm table for numbers between 0 and 10 at the end of this book, pages 74 and 75.
This is a three-figure table and will only give answers to that degree of accuracy.

Use the table. Find the logarithm of each of the following numbers. Write each number to base 10.

(1) 7·53   (2) 1·46   (3) 3·38   (4) 7·09   (5) 5·85

(6) 3·3   (7) 6·72   (8) 7·04   (9) 25·6   (10) 83·1

(11) 79·4   (12) 50   (13) 500   (14) 74·6   (15) 746

B

In this exercise you are given the logarithm of a number.
In each case find the number.

[You will notice that in some cases two numbers have the same logarithm. This is due to the adjustment to 3 figures and either answer will be considered correct.]

(1) 0·525   (2) 0·253   (3) 0·741   (4) 0·487   (5) 0·574
(6) 0·377   (7) 0·916   (8) 1·386   (9) 1·179   (10) 1·847
(11) 0·561   (12) 1·561   (13) 2·561   (14) 2·906   (15) 2·962

Using Logarithms — Multiplication

Example $6·47 \times 3·82$
$= 10^{0·811} \times 10^{0·582} = 10^{(0·811 + 0·582)}$
$= 10^{1·393}$
$= 24·7$

Complete each of the following calculations using the logarithm table.

(1) 2·52 × 1·87   (2) 3·03 × 2·16   (3) 1·44 × 1·93
(4) 2·85 × 2·13   (5) 7·46 × 1·26   (6) 8·53 × 3·57
(7) 17·6 × 2·78   (8) 46·7 × 3·08   (9) 21·8 × 9·47
(10) 64·8 × 4·34   (11) 17·4 × 13·7   (12) 36·9 × 23·4

Using Logarithms — Division

Example $86·3 \div 9·57$
$= 10^{1·936} \div 10^{0·981}$
$= 10^{(1·936 - 0·981)}$
$= 10^{0·955}$
$= 9·02$

Complete each of the following calculations using the logarithm table.

(1) 3·94 ÷ 1·85   (2) 5·48 ÷ 3·16   (3) 8·93 ÷ 6·92
(4) 9·09 ÷ 3·56   (5) 7·46 ÷ 4·34   (6) 24·3 ÷ 7·87
(7) 89·4 ÷ 9·52   (8) 95·7 ÷ 37·2   (9) 697 ÷ 43·3
(10) 845 ÷ 6·63   (11) 246 ÷ 7·03   (12) 100 ÷ 7·84

Finding Volume

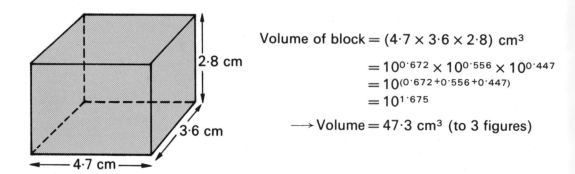

Volume of block = $(4{\cdot}7 \times 3{\cdot}6 \times 2{\cdot}8)$ cm^3

$= 10^{0{\cdot}672} \times 10^{0{\cdot}556} \times 10^{0{\cdot}447}$
$= 10^{(0{\cdot}672 + 0{\cdot}556 + 0{\cdot}447)}$
$= 10^{1{\cdot}675}$

$\longrightarrow$ Volume $= 47{\cdot}3$ cm^3 (to 3 figures)

E

Use logarithms to find the volume of each of the following solids.
(The diagrams are not drawn to scale.)

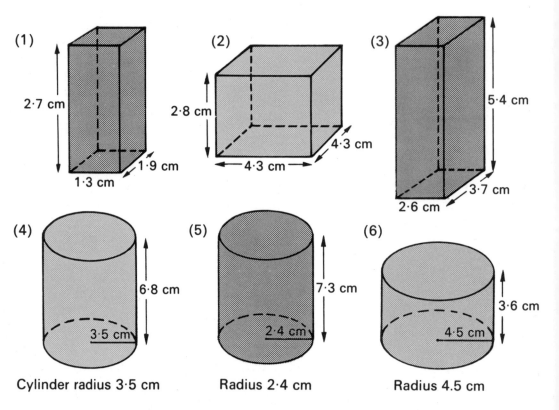

(4) Cylinder radius 3·5 cm

(5) Radius 2·4 cm

(6) Radius 4.5 cm

Take the value of π to be 3·14.

Polyhedra

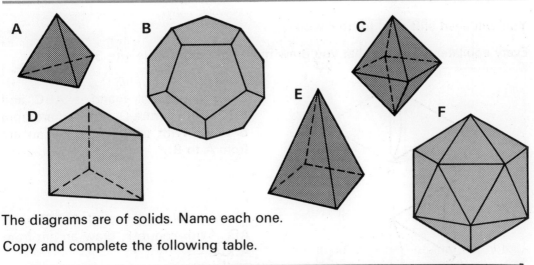

The diagrams are of solids. Name each one.

Copy and complete the following table.

	Polyhedron	No. of faces F	No. of vertices V	No. of edges E	(F+V)	(F+V)−E
(1)	Tetrahedron					
(2)	Cube					
(3)	Octahedron					
(4)	Dodecahedron					
(5)	Icosahedron					
(6)	Triangular prism					
(7)	Rectangular prism (Cuboid)					
(8)	Right pyramid on square base					

You will notice that the sum of the number of faces and the number of vertices is always 2 more than the number of edges. This relation was suggested by a famous Swiss mathematician called Euler, who lived in the eighteenth century.

Tessellation

You will need stiff card for this work.

Every equilateral triangle that you draw must have sides of length 3 cm.

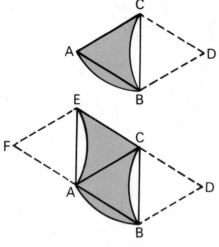

Draw equilateral triangles ABC and BCD. With centre D, draw an arc from B to C. With centre C, draw an arc from A to B.

Draw equilateral triangles ACE and AEF. With centre F, draw an arc from A to E.

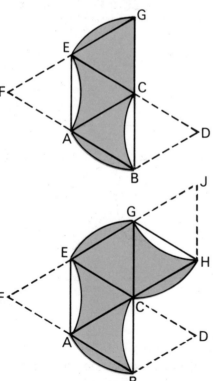

Draw equilateral triangle CEG. With centre C, draw an arc from E to G.

Draw equilateral triangles CGH and GHJ. With centre G, draw an arc from C to H. With centre J, draw an arc from G to H.

Cut out the coloured shape.

Use it to make this pattern.

Income Tax

The rate of Income Tax for the current year is announced by the Chancellor of the Exchequer in his Budget Speech.
Income Tax is not paid on the whole income—allowances are made on which no tax is paid.
Income Tax for the year 1971–1972 was 38·75% of taxable income.
There are higher rates of taxation but they will not concern us.

Example
Find the tax paid in 1 year by Mr White.

Annual income ——— £1200
Allowances ——— £600

Taxable income ——— £600

Tax paid = 38·75% of £600
= £(38·75 × 6)
= £232·50

Find the amount of Income Tax paid in each of the following cases at the rate quoted above.

	Name	Annual Income	Allowances
(1)	Jane Adams	£900	£500
(2)	David Bell	£1760	£860
(3)	Sheila Croft	£1120	£520
(4)	Richard Davies	£960	£1030
(5)	Peter Ford	£1270	£1070
(6)	Allan Moore	£2110	£1010
(7)	Robert Pearce	£1530	£1560
(8)	Carol Reid	£1380	£680
(9)	Joan Smith	£1760	£960
(10)	John Thomas	£1820	£1520
(11)	Julie White	£1240	£740
(12)	Martin Young	£1050	£950

Find the amount of Income Tax that would be paid by each of the people named in **A** at the time you are doing this work.

Income Tax Allowances

You will have noticed that allowances vary, so that people in the higher income groups do not always pay the most tax. The reasons for this are given below.

Tax Allowances in 1971–1972

Earned income —— An allowance of $\frac{2}{9}$ of net earned income
Single person —— A personal allowance of £325
Married person —— A personal allowance of £465

Allowance for children
- For every child under 11 years £155
- For every child between 11 and 16 years £180
- For every child over 16 years who is at school or college £205

Family Allowance

There is no allowance for a first child, 90p per week for a second child, and £1 per week for every other child. These allowances are added to the earned income to find the total income for tax purposes.

Example

Find the taxable income for Mr Grant, whose annual earned income is £1800. He has 3 children under 11 years.

Earned income —— £1800
less —— £ 400 —— $\frac{2}{9}$ of earned income
£1400
less —— £ 465 —— Married man's allowance
£ 935
less —— £ 465 —— Children's allowance—3 at £155
£ 470
plus —— £ 98·8 —— Family allowance—1 at £46·80, 1 at £52

Taxable income —— £ 568·8

Find the taxable income in each of the following cases.

	Name	Married or single	Children	Annual earned income
(1)	Bob Croft	Single	—	£900
(2)	Paul Davies	Married	1 under 11 yr.	£1800
(3)	Richard Eaton	Married	Twins 2 yr.	£1350
(4)	Christine Gates	Single	—	£1080
(5)	Stephen Jones	Married	1 of 13 yr. 1 of 17 yr. at school	£2160
(6)	Ian Kerr	Married	1 of 9 yr. 1 of 13 yr.	£1710
(7)	Gill Long	Single	—	£1440
(8)	Jeremy Pike	Married	1 of 3 yr. 1 of 12 yr. 1 of 18 yr. at school	£2340

Drawing to Scale

(The ground must be considered horizontal in each of these questions and the diagrams are not drawn to scale.)

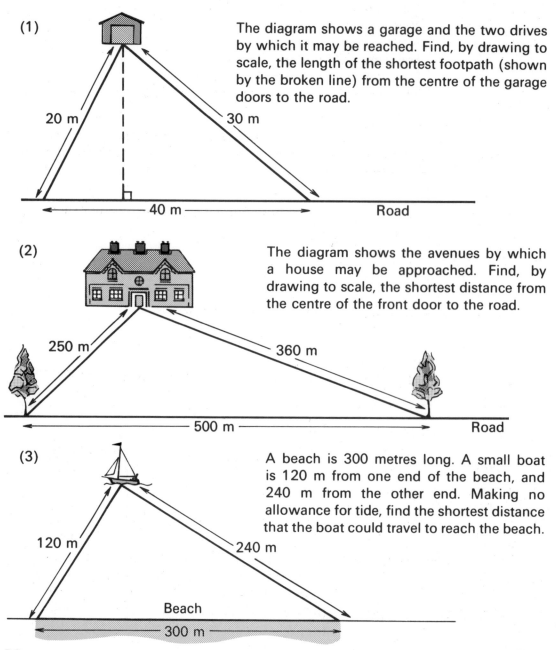

(1) The diagram shows a garage and the two drives by which it may be reached. Find, by drawing to scale, the length of the shortest footpath (shown by the broken line) from the centre of the garage doors to the road.

(2) The diagram shows the avenues by which a house may be approached. Find, by drawing to scale, the shortest distance from the centre of the front door to the road.

(3) A beach is 300 metres long. A small boat is 120 m from one end of the beach, and 240 m from the other end. Making no allowance for tide, find the shortest distance that the boat could travel to reach the beach.

(4) 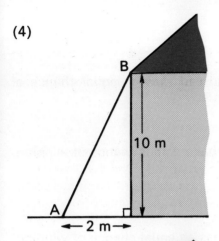 Find the length of the shortest ladder which would reach B, the edge of the roof of a house, from a point A on the ground, 2 m from the wall of the house.

(5) 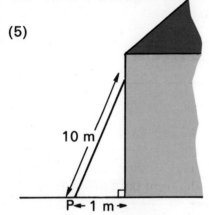 Find the greatest height which could be reached by a ladder 10 m long, from a point P on the ground 1 m from the wall of the house.

(6) 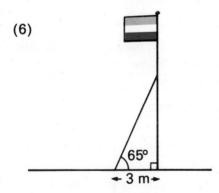 One of the wire stays which supports a flagpole is fixed at a distance of 3 m from the foot of the flagpole. The wire stay makes an angle of 65° with the ground.
(a) What is the length of the wire stay?
(b) How high up the flagpole is the stay attached?

(7) 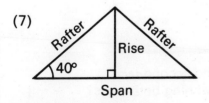 The diagram shows two of the rafters of a roof. Each rafter makes an angle of 40° with the horizontal. The span is 9 m.
(a) What is the rise?
(b) What is the length of each rafter?

Probability

Three runners P, Q and R take part in a race and each has an equal chance of winning.

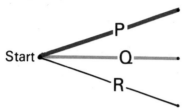

Each has a 1 in 3 chance of winning.

They then take part in a second race, again each has an equal chance of winning. The tree chart shows the possible results of the two races.

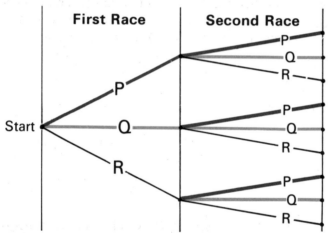

The diagram shows that if P wins the first race, then P, Q or R may win the second. Similarly if Q wins the first race, then P, Q or R may win the second.
If R wins the first race then P, Q or R may win the second.

(1) Copy and complete this table showing all the outcomes.

Winner Race 1	Winner Race 2
P $\begin{cases} P \\ P \\ P \end{cases}$	P Q R

(2) What is the chance that P will win both first and second race?
(3) Which other competitors have the chance of winning both races?

If the 3 runners, P, Q and R, then take part in a third race what chance has Q of winning all three races?

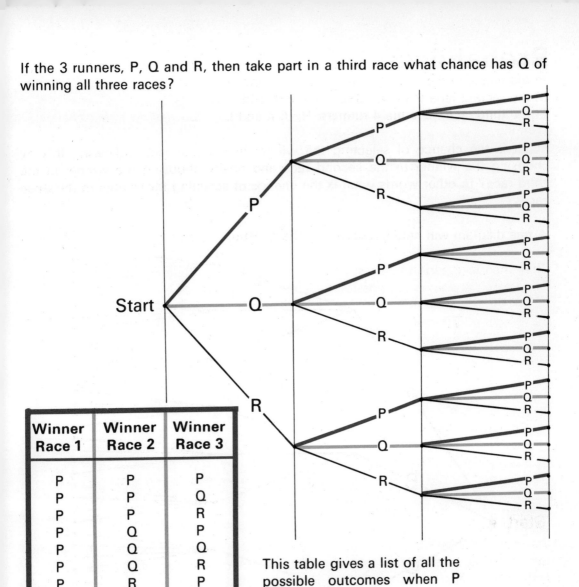

Winner Race 1	Winner Race 2	Winner Race 3
P	P	P
P	P	Q
P	P	R
P	Q	P
P	Q	Q
P	Q	R
P	R	P
P	R	Q
P	R	R

This table gives a list of all the possible outcomes when P wins the first race.

(1) Make a table showing all the possible outcomes when Q wins the first race.
(2) Make a table showing all the possible outcomes when R wins the first race.
(3) What is the chance that P will win all three races?
(4) What is the chance that Q will win all three races?
(5) What is the chance that R will win all three races?
(6) Study the tree chart carefully and say what chance P will have of winning two races only.
(7) What is the chance that R will win one race only?
(8) What is the chance that Q will win no race?

53

Three races are to be run.
In the first race there are 4 runners, A, B, C and D.
In the second race there are 3 runners, E, F and G.
In the third race there are 4 runners, H, J, K and L.

What is the chance of selecting the winner in the first race, following this by choosing the winner in the second race and finally choosing the winner in the third race? In other words what is the chance of selecting the winner in the three races?

A tree diagram will help in answering the question.

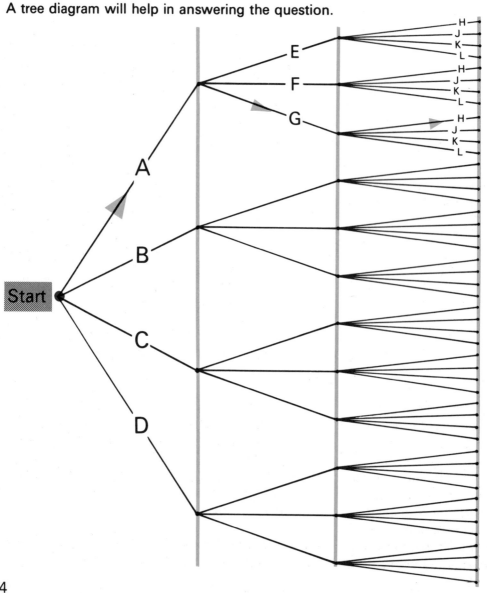

The diagram shows the possibilities when A wins the first race. Then E, F or G might win the second race. If A wins the first race and E wins the second, then H or J or K or L may win the third race.

Similarly if A wins the first and F wins the second, then H, J, K or L may win the third race. If A wins the first and G wins the second, then H, J, K or L may win the third.
But, of course, B or C or D may win the first race.

The table below gives all the possible outcomes when A wins the first race, as shown in the tree diagram.

Winner race 1	Winner race 2	Winner race 3
A	E	H
A	E	J
A	E	K
A	E	L
A	F	H
A	F	J
A	F	K
A	F	L
A	G	H
A	G	J
A	G	K
A	G	L

(1) Take a large sheet of paper, copy the tree diagram as shown on the opposite page and then complete it to show what may happen: (a) if B wins the first race; (b) if C wins the first race; (c) if D wins the first race.
(2) Make a table similar to that shown above showing all the outcomes when (a) B wins the first race; (b) C wins the first race; (c) D wins the first race.
(3) How many outcomes are possible with the given runners in the three races?
(4) The yellow arrow on the tree diagram shows that A wins the first race, G wins the second race and H wins the third race.
What is the chance of this happening?
(5) Choose a winner in each race. What chance have you of choosing the three winners? Show this with coloured arrows on your tree diagram.

When a coin is tossed it may fall with head uppermost or tail uppermost. Suppose you call 'Heads'. Do you think you have a better chance of being right if one coin is tossed or if two coins are tossed? Write down your answer before doing any more work on the subject.

A tree diagram helps.

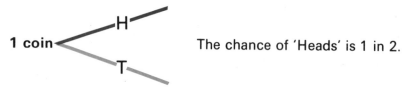

2 coins In this case it is easier to choose two different coins. They may fall as shown below.

This may be shown in a tree diagram.

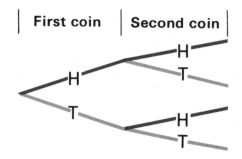

The chance of both coins falling with 'Head' uppermost is 1 in 4. Was your answer correct?

The diagram shows what may happen, that is all the outcomes, when three different coins are tossed.

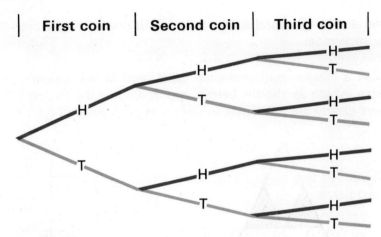

The diagram shows that the coins may fall in the following ways.

```
3 Heads:──── H  H  H      1 way
             ⎧ H  H  T
2 Heads:──── ⎨ H  T  H    3 ways
             ⎩ T  H  H
             ⎧ H  T  T
1 Head: ──── ⎨ T  H  T    3 ways
             ⎩ T  T  H
No Heads:─── T  T  T      1 way
```

This shows clearly that there are eight outcomes. The chance of calling 'Heads' and being correct for three coins is 1 in 8.

---

(1) Can you suggest a reason why three 'different' coins are used?

(2) **Using four different coins**: Make a tree diagram showing all the ways in which four coins may fall.

(3) Write down the number of ways in which the four coins may fall with
    (a) 4 heads   (b) 3 heads   (c) 2 heads   (d) 1 head   (e) no heads
    uppermost.

(4) What is the chance of four heads being uppermost when four coins are tossed?

(5) What is the chance of heads being uppermost on all the coins when (a) 5 (b) 7 (c) 8 coins are tossed?

Pascal's Triangle

When a coin is tossed it can only fall in one of two ways—heads uppermost or tails uppermost.

Pascal, a French mathematician who lived in the seventeenth century, drew the pattern which is shown below and is known as Pascal's Triangle. The pattern shows clearly the outcomes when coins are tossed.

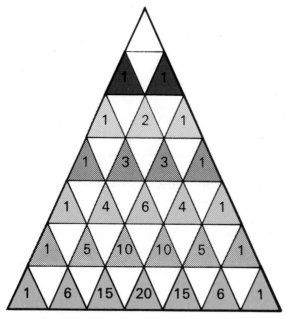

	Outcomes	Chance		
1 coin	2	1 head $\frac{1}{2}$	0 head $\frac{1}{2}$	
2 coins	4	2 heads $\frac{1}{4}$	1 head $\frac{2}{4}$	0 heads $\frac{1}{4}$
3 coins	8	3 heads $\frac{1}{8}$ 0 heads $\frac{1}{8}$	2 heads $\frac{3}{8}$	1 head $\frac{3}{8}$
4 coins	16	4 heads $\frac{1}{16}$ 1 head $\frac{4}{16}$	3 heads $\frac{4}{16}$ 0 heads $\frac{1}{16}$	2 heads $\frac{6}{16}$
5 coins	32	5 heads $\frac{1}{32}$ 2 heads $\frac{10}{32}$	4 heads $\frac{5}{32}$ 1 head $\frac{5}{32}$	3 heads $\frac{10}{32}$ 0 heads $\frac{1}{32}$
6 coins	64	6 heads $\frac{1}{64}$ 3 heads $\frac{20}{64}$ 0 heads $\frac{1}{64}$	5 heads $\frac{6}{64}$ 2 heads $\frac{15}{64}$	4 heads $\frac{15}{64}$ 1 head $\frac{6}{64}$

(1) Use the triangle to check your answer to questions (5) (a) on the previous page.
(2) Write down the line of the pattern when seven coins are tossed.
(3) What is the number of outcomes when seven coins are tossed?
(4) What is the chance that (a) 7 heads   (b) 6 heads   (c) 5 heads   (d) 4 heads   (e) 3 heads   (f) 2 heads   (g) 1 head   (h) no head fall uppermost?
(5) Check your answer to (5) (b) on the previous page.
(6) Write down the line of the pattern when eight coins are tossed.
(7) What is the number of outcomes when eight coins are tossed?
(8) What is the chance that (a) 8 heads   (b) 7 heads   (c) 6 heads   (d) 5 heads   (e) 4 heads   (f) 3 heads   (g) 2 heads   (h) 1 head   (i) no head fall uppermost?
(9) Check your answer to question (5) (c) on the previous page.

Probability with Dice

When an ordinary six-faced die is thrown the score on the face uppermost may be 1, 2, 3, 4, 5 or 6. When a die is thrown you have a one in six chance of scoring a particular number between 1 and 6.

Two dice are thrown and the two scores are added.

The table shows the possible total scores.

Score on yellow die

Score on violet die

	1	2	3	4	5	6
1						
2						
3				7		
4						
5						11
6					11	

When 3 is scored on the yellow die and 4 on the violet die the total score is 7. This is shown in the table.
When 5 is scored on the yellow die and 6 on the violet die the total score is 11. Again this is shown.
11 could also be scored by getting 6 on the yellow die and 5 on the violet die.

(1) Copy and complete the table to show all the total scores.

(2) Use squared paper and make a column graph showing the frequency of each total score.

The frequency for scoring 11 is shown.

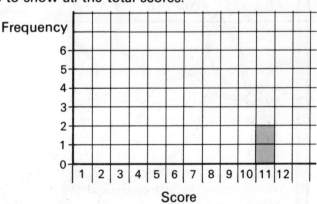

(3) When you throw two dice what is the chance that you will score (a) 12 (b) 10 (c) 8 (d) 7 (e) 5?
(4) Can you give any reason for the difference in the frequency of the various scores?

A Tour of the Wye Valley

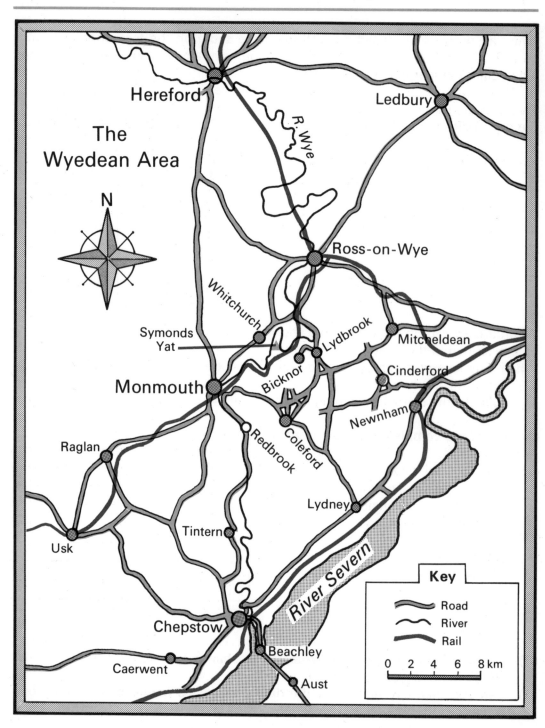

The party stayed at Ross-on-Wye and returned there each evening.

A

(1) In which direction, as the crow flies, is each of the following towns from Ross-on-Wye? Give the bearing in each case.

(a) Hereford   (b) Monmouth   (c) Chepstow   (d) Ledbury   (e) Usk
(f) Cinderford   (g) Tintern   (h) Lydney   (j) Coleford   (k) Raglan

(2) Find the distance, as the crow flies, of each of the above towns from Ross-on-Wye. Give your answers to the nearest kilometre.

(3) Put a dot in the centre of your paper to represent Ross. Drawing carefully and to scale, show the positions and distances of the ten towns from Ross.

B

Find the shortest distance travelling by road between:

(1) Ross and Hereford
(2) Ross and Monmouth
(3) Ross and Chepstow
(4) Ross and Usk
(5) Ross and Ledbury
(6) Monmouth and Chepstow
(7) Hereford and Ledbury
(8) Tintern and Cinderford
(9) Monmouth via Whitchurch and Lydbrook to Symonds Yat and by the quickest route back to Monmouth (circular tour)

C

(1) Find the length of the River Wye from where the railway crosses the river just south of Hereford, to where the railway crosses the river a little to the west of Ross.
(2) Find the difference in distance travelled between Hereford and Ross by road and by river.
(3) Find the length of the River Wye between Monmouth and Chepstow Bridge.
(4) If you could travel by the river between these two places, how much farther would you have gone than the distance by road?
(5) Find the length of the bank of the River Severn on its western side between Newnham and the Aust-Beachley Bridge.

Savings

When you have saved some money you will need to put it where it will be safe and also earn some interest.

There are many ways of investing money. Below are details of four different ways in which money can be saved.
Each of those shown below has advantages and disadvantages.

A	B	C	D
National Savings Bank Ordinary Account	**National Savings Bank Investment Account**	**Bank— Deposit Account**	**Building Society**
Interest $3\frac{1}{2}$% p.a. Tax paid on amounts of interest above £21 p.a.	Interest $7\frac{1}{2}$% p.a. £50 must be held in National Savings Bank Ordinary Account	Interest 5% p.a. Interest taxed at current standard rate	Interest $4\frac{1}{2}$% p.a. Interest is tax free as the tax has been paid before interest is received

An investor needs to compare the interest he will gain on his money. He does not need to know the exact amount before deciding which savings plan is best for the amount of money he has to invest.

The standard rate of income tax in the year in which this book is written is 38·75%. This means that on every £1 of taxable income 38·75p will be deducted so that for every £1 the net income is $(100 - 38 \cdot 75)p = 61 \cdot 25p$.
For estimating, this can be adjusted to the nearest penny—61p.
The actual amount received should be slightly more than the estimate.

Example Jack Boon has saved £200 which he wishes to invest for 1 year. He is paying Income Tax at the standard rate. He compares each of the four investments shown above.

A   On £200 for 1 year he will receive £7 on which he pays no tax.

B   He has to put £50 in Ordinary Account and gets £1·75 tax free. This leaves £150 to put in National Savings Bank Investment Account. His gross interest is £$(7\frac{1}{2} \times 1\frac{1}{2})$ = £11·25, but after tax is deducted the net interest is $(11\cdot25 \times 61)$p = £6·86 (to nearest p). Total interest is £6·86 + £1·75 = £8·61.

C   On £200 for 1 year he receives gross interest of £10. After tax deduction his net interest is (10×61)p = £6·10.

D   Interest, tax already paid = £$(4\frac{1}{2} \times 2)$ = £9.

If there is no special reason to invest in one rather than the other, then he gains most from **D**.

These rates of interest were applicable when this book was written. Compare these rates with those paid now.

Decide which method of saving is possible for each of the following amounts. Compare the net interest on each method and decide which is the best investment for the amount under consideration.

(1)   Amount to be invested   £50

(2)   Amount to be invested   £100

(3)   Amount to be invested   £250

(4)   Amount to be invested   £40

(5)   Amount to be invested £1000

(6)   Amount to be invested £1750

Family Tree

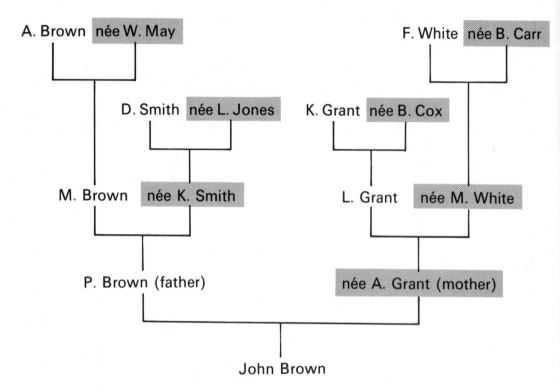

(1) Copy and complete:
John had ---- parents.
He had ---- grandparents.
He had ---- great-grandparents.

(2) How many great-great-grandparents did John have?

(3) How many great-great-great-grandparents did John have?

(4) Which surname survives through the four generations?

(5) Deborah was John's sister. How many great-grandparents did she have?

(6) When John marries and has a son, what will his surname be?

(7) Write down the names of the baby's grandparents on his father's side.

(8) What will be the surname of Deborah's son?

Ancestry of a Drone Bee

A male bee, called a drone, has a mother but no father. If eggs laid by a female bee are fertilized, they hatch into females, either queens or workers. Males hatch from unfertilized eggs.

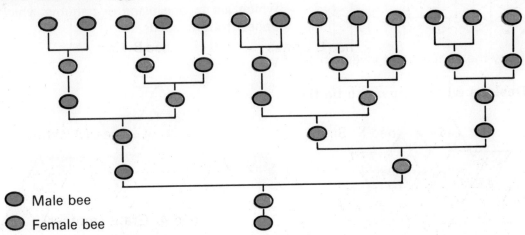

○ Male bee
○ Female bee

For the sake of clarity, we will refer to the drone's parents, grandparents, etc. Copy and complete:

		Number of Female	Number of Male	Total
(1)	parents	1	0	1
(2)	grandparents	1	1	2
(3)	great-grandparents			
(4)	great-great-grandparents			
(5)	great-great-great-grandparents			
(6)	great-great-great-great-grandparents			

Copy the diagram and add the next generation to the family tree of the drone bee.
The sequence of numbers 1, 1, 2, 3, 5, 8, 13 . . . is called the Fibonacci Sequence.

Using Isometric Paper

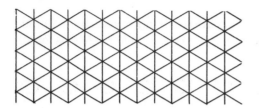

This paper is ruled into equilateral triangles and is said to form an isometric grid.
The paper can be used for drawing patterns which are in one plane or patterns which appear to form solids.

Copy the patterns and invent some of your own.

Designs which appear to be flat

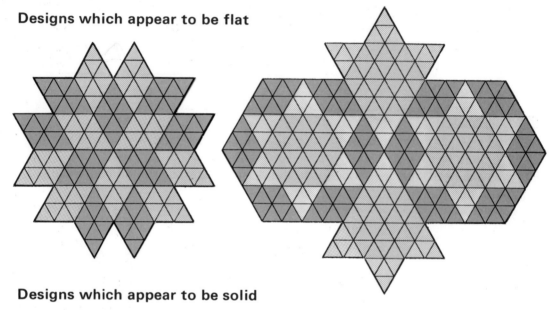

Designs which appear to be solid

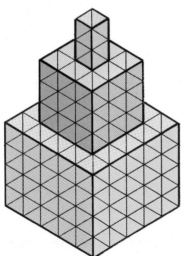
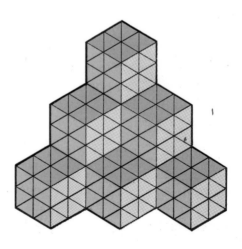

Reasoning

In each of the following examples the statements labelled **X** and **Y** must be accepted as being true.

From the two statements it is sometimes possible to deduce another fact but care must be taken to avoid faulty reasoning.

Use the statements **X** and **Y** to answer the question that follows in each of the following.

(1) **X** | Form VA won the football cup in a knock-out competition.
     Y | The final match was between VA and VI B. Form VA scored one goal in this match.
How many goals were scored by VI B in the final?

(2) **X** | Michael cycles at a greater speed than that at which his brother Ian walks.
     Y | Both set out from home at the same time and neither stopped on their way to school.
One boy was detained because he arrived at school late. Which boy was detained?

(3) **X** | All men and women over the age of 18 can vote in a general election.
     Y | Philip White is 21.
Could Philip vote in the General Election?

(4) **X** | Norwich is in the county of Norfolk.
     Y | Frank lives in Norfolk.
Does Frank live in Norwich?

(5) **X** | All the senior prefects are in the Sixth Form.
     Y | Ted Forbes is a senior prefect.
Is Ted in the Sixth Form?

(6) **X** | All senior prefects are in the Sixth Form.
     Y | Vera is in the Sixth Form.
Is Vera a Senior Prefect?

(7) **X** | $a < b$
     Y | $b < c$
Is c less than a?

(8) **X** | $m > k$
     Y | $n > k$
Is m greater than n?

Compound Interest

When money is invested for a period longer than one year compound interest is paid. This means that the interest gained in the first year is added to the original principal. This increased sum becomes the principal for the second year. Again the interest gained in the second year is added and the increased sum becomes the principal for the third year. This process is repeated as long as the interest is not withdrawn.

Follow this example carefully.
£500 is invested for 3 years at a 4% per year rate of interest.
Find, to the nearest £0·01: (a) the amount at the end of 3 years;
(b) the interest gained in 3 years.

	£	
Original principal	= 500	
Interest for first year	= 20	— 4% on £500
Principal for second year	= 520	
Interest for second year	= 20·80	— 4% of £520
Principal for third year	= 540·80	
Interest for third year	= 21·632	— 4% on £540·8
Amount after three years	= £562·432	

(a) The amount at the end of 3 years = £562·43 (to nearest £0·01)
(b) Interest gained in 3 years = £562·43 − £500 = £62·43

In each of the following find, to the nearest £0·01,
(a) the amount at the end of the time stated,
(b) the interest gained in the time under consideration.

	Original principal	Time under consideration	Rate of interest per year
(1)	£800	3 years	5%
(2)	£300	2 years	4%
(3)	£600	2 years	6%
(4)	£250	2 years	6%
(5)	£1000	3 years	5%
(6)	£150	3 years	4%
(7)	£2000	4 years	5%
(8)	£700	3 years	7%
(9)	£450	3 years	4%
(10)	£50	4 years	6%

Depreciation

A machine was bought for £750. Its value depreciated by 10% per year. What was its value at the end of the second year?

    Buying price   = £750
    Depreciation in first year   = £75 ———— 10% of £750

    Value at end of first year   = £675
    Depreciation in second year = £67·50 ———— 10% of £675

    Value at end of second year = £607·50

        Value after 2 years   = £607·50

        Depreciation in 2 years = £750 − £607·50 = £142·50

Find, to the nearest £0·01, (a) the value at the end of the time stated,
(b) the amount of the depreciation,
in each of the following.

		Original value	Depreciation per year	Number of years
(1)	Washing machine	£100	10%	2 years
(2)	Bicycle	£30	10%	2 years
(3)	Van	£1250	20%	2 years
(4)	Lorry	£2500	20%	2 years
(5)	Car	£1000	25%	2 years
(6)	Office equipment	£5500	10%	3 years
(7)	Electric equipment	£6000	20%	2 years
(8)	Factory machinery	£10,000	15%	2 years
(9)	Garage equipment	£8000	5%	3 years
(10)	Rolling stock	£100,000	20%	3 years

The Nephroid

P———————Q   Draw a line PQ 4 cm long.

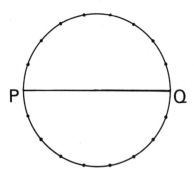

Draw a circle with diameter PQ. This circle is called the base circle.

Mark a number of equally spaced points on the circumference of the base circle. (16 points have been marked in this case.)

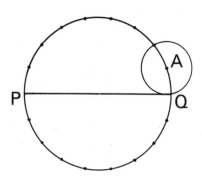

With centre A, draw a circle which just touches the line PQ.

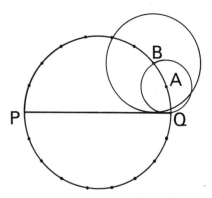

With centre B, draw a circle which just touches the line PQ.

With each point marked on the base circle as centre, draw a circle which just touches the line PQ.

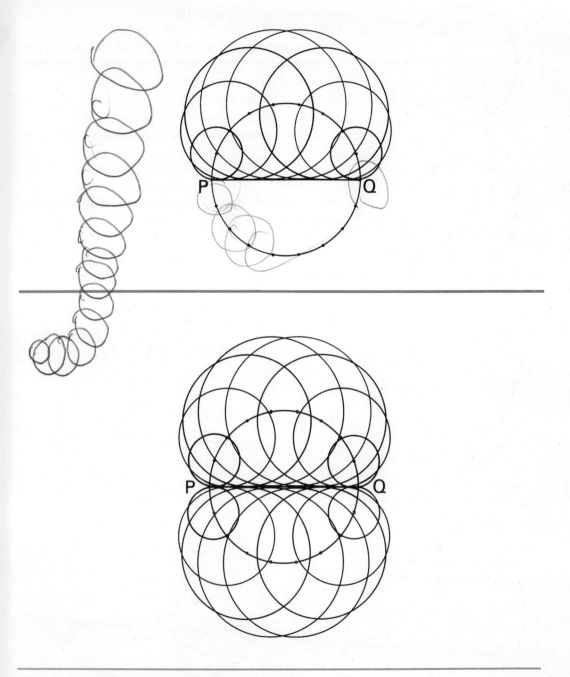

The word **nephroid** means 'kidney-shaped'.

The Cardioid

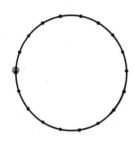

Draw a circle of radius 1½ cm.
This circle is called the base circle.
Mark a number of equally spaced points on the circumference of the base circle. (18 points have been marked in this case.)
Mark one of the points in blue.

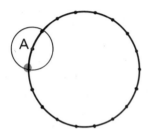

With centre A, draw a circle whose circumference passes through the blue point.

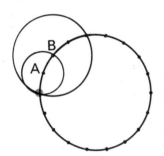

With centre B, draw a circle whose circumference passes through the blue point.

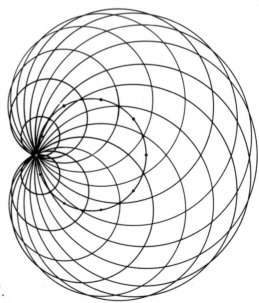

With each point on the base circle as centre, draw a circle whose circumference passes through the blue point.

The word **cardioid** means 'heart-shaped'.

The Spiral

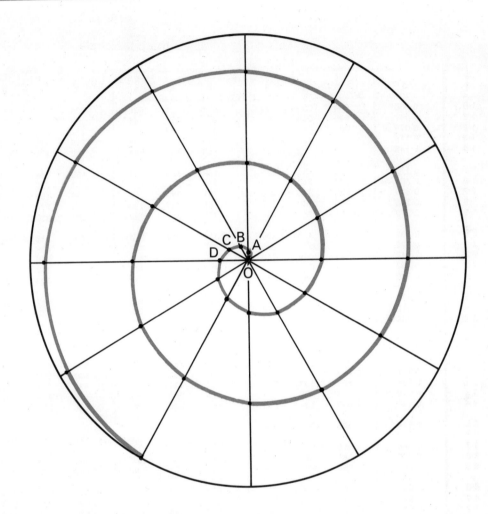

Draw a circle, centre O, of radius 6 cm.

Mark angles of 30° at the centre, and draw the radii of the circle as shown.

On one of the radii, mark a point A such that OA = 0·2 cm.
On the next radius, mark a point B such that OB = 0·4 cm.
On the next radius, mark a point C such that OC = 0·6 cm.
Continue in this way, so that each point is 0·2 cm further from O than the preceding point.

Join OABC . . . by a curved line to form the spiral.

Logarithm Tables

	0	1	2	3	4	5	6	7	8	9
1·0	0·000	004	009	013	017	021	025	029	033	037
1·1	0·041	045	049	053	057	061	064	068	072	076
1·2	0·079	083	086	090	093	097	100	104	107	111
1·3	0·114	117	121	124	127	130	134	137	140	143
1·4	0·146	149	152	155	158	161	164	167	170	173
1·5	0·176	179	182	185	188	190	193	196	199	201
1·6	0·204	207	210	212	215	217	220	223	225	228
1·7	0·230	233	236	238	241	243	246	248	250	253
1·8	0·255	258	260	262	265	267	270	272	274	276
1·9	0·279	281	283	286	288	290	292	294	297	299
2·0	0·301	303	305	307	310	312	314	316	318	320
2·1	0·322	324	326	328	330	332	334	336	338	340
2·2	0·342	344	346	348	350	352	354	356	258	360
2·3	0·362	364	365	367	369	371	373	375	377	378
2·4	0·380	382	384	386	387	389	391	393	394	396
2·5	0·398	400	401	403	405	407	408	410	412	413
2·6	0·415	417	418	420	422	423	425	427	428	430
2·7	0·431	433	435	436	438	439	441	442	444	446
2·8	0·447	449	450	452	453	455	456	458	459	461
2·9	0·462	464	465	467	468	470	471	473	474	476
3·0	0·477	479	480	481	483	484	486	487	489	490
3·1	0·491	493	494	496	497	498	500	501	502	504
3·2	0·505	507	508	509	511	512	513	515	516	517
3·3	0·519	520	521	522	524	525	526	528	529	530
3·4	0·531	533	534	535	537	538	539	540	542	543
3·5	0·544	545	547	548	549	550	551	553	554	555
3·6	0·556	558	559	560	561	562	563	565	566	567
3·7	0·568	569	571	572	573	574	575	576	577	579
3·8	0·580	581	582	583	584	585	587	588	589	590
3·9	0·591	592	593	594	595	597	598	599	600	601
4·0	0·602	603	604	605	606	607	609	610	611	612
4·1	0·613	614	615	616	617	618	619	620	621	622
4·2	0·623	624	625	626	627	628	629	630	631	632
4·3	0·633	634	635	636	637	638	639	640	641	642
4·4	0·643	644	645	646	647	648	649	650	651	652
4·5	0·653	654	655	656	657	658	659	660	661	662
4·6	0·663	664	665	666	667	667	668	669	670	671
4·7	0·672	673	674	675	676	677	678	679	679	680
4·8	0·681	682	683	684	685	686	687	688	688	689
4·9	0·690	691	692	693	694	695	695	696	697	698
5·0	0·699	700	701	702	702	703	704	705	706	707
5·1	0·708	708	709	710	711	712	713	713	714	715
5·2	0·716	717	718	719	719	720	721	722	723	723
5·3	0·724	725	726	727	728	728	729	730	731	732
5·4	0·732	733	734	735	736	736	737	738	739	740

	0	1	2	3	4	5	6	7	8	9
5·5	0·740	741	742	743	744	744	745	746	747	747
5·6	0·748	749	750	751	751	752	753	754	754	755
5·7	0·756	757	757	758	759	760	760	761	762	763
5·8	0·763	764	765	766	766	767	768	769	769	770
5·9	0·771	772	772	773	774	775	775	776	777	777
6·0	0·778	779	780	780	781	782	782	783	784	785
6·1	0·785	786	787	787	788	789	790	790	791	792
6·2	0·792	793	794	794	795	796	797	797	798	799
6·3	0·799	800	801	801	802	803	803	804	805	806
6·4	0·806	807	808	808	809	810	810	811	812	812
6·5	0·813	814	814	815	816	816	817	818	818	819
6·6	0·820	820	821	822	822	823	823	824	825	825
6·7	0·826	827	827	828	829	829	830	831	831	832
6·8	0·833	833	834	834	835	836	836	837	838	838
6·9	0·839	839	840	841	841	842	843	843	844	844
7·0	0·845	846	846	847	848	848	849	849	850	851
7·1	0·851	852	852	853	854	854	855	856	856	857
7·2	0·857	858	859	859	860	860	861	862	862	863
7·3	0·863	864	865	865	866	866	867	867	868	869
7·4	0·869	870	870	871	872	872	873	873	874	874
7·5	0·875	876	876	877	877	878	879	879	880	880
7·6	0·881	881	882	883	883	884	884	885	885	886
7·7	0·886	887	888	888	889	889	890	890	891	892
7·8	0·892	893	893	894	894	895	895	896	897	897
7·9	0·898	898	899	899	900	900	901	901	902	903
8·0	0·903	904	904	905	905	906	906	907	907	908
8·1	0·908	909	910	910	911	911	912	912	913	913
8·2	0·914	914	915	915	916	916	917	918	918	919
8·3	0·919	920	920	921	921	922	922	923	923	924
8·4	0·924	925	925	926	926	927	927	928	928	929
8·5	0·929	930	930	931	931	932	932	933	933	934
8·6	0·934	935	936	936	937	937	938	938	939	939
8·7	0·940	940	941	941	942	942	943	943	943	944
8·8	0·944	945	945	946	946	947	947	948	948	949
8·9	0·949	950	950	951	951	952	952	953	953	954
9·0	0·954	955	955	956	956	957	957	958	958	959
9·1	0·959	960	960	960	961	961	962	962	963	963
9·2	0·964	964	965	965	966	966	967	967	968	968
9·3	0·968	969	969	970	970	971	971	972	972	973
9·4	0·973	974	974	975	975	975	976	976	977	977
9·5	0·978	978	979	979	980	980	980	981	981	982
9·6	0·982	983	983	984	984	985	985	985	986	986
9·7	0·987	987	988	988	989	989	989	990	990	991
9·8	0·991	992	992	993	993	993	994	994	995	995
9·9	0·996	996	997	997	997	998	998	999	999	1·000
10·0	1·000									

Useful Information

Foreign Currency and Exchange Rates

The rates of exchange given below are those for the year in which this book was written (1971). As you know, the rate of exchange may vary from day to day. Should you need to know the exact rate you can obtain it from a bank or you may be able to get the information from a daily paper.

Country	Currency	Rate of Exchange
Austria	Schilling 100 groschen = 1 schilling	£1 = 62·40 S.
Belgium	Franc 100 centimes = 1 franc	£1 = 120 B.Fr.
Denmark	Krone 100 öre = 1 krone	£1 = 18 D.Kr.
Finland	Markka 100 pennia = 1 markka	£1 = 10·8 F.Mk.
France	Franc 100 centimes = 1 franc	£1 = 13·35 Fr.
West Germany	Deutsche Mark 100 pfennig = 1 mark	£1 = 8·60 D.M.
Greece	Drachma 100 lepta = 1 drachma	£1 = 72 Dr.
Italy	Lire	£1 = 1505 I.
Netherlands	Florin (Guilder) 100 cents = 1 guilder	£1 = 8·60 G.
Norway	Krone 100 öre = 1 krone	£1 = 17·18 N.Kr.
Portugal	Escudo 100 centavos = 1 escudo	£1 = 69 Esc.
Spain	Peseta 100 céntimos = 1 peseta	£1 = 168 Ptas.
Sweden	Krona 100 öre = 1 krona	£1 = 12·50 S.Kr.
Switzerland	Franc 100 centimes = 1 franc	£1 = 9·95 Sw.Fr.
Yugoslavia	Dinar 100 paras = 1 dinar	£1 = 30 din.
U.S.A.	Dollar 100 cents = 1 dollar	£1 = $2·40